Human Factors in Challenging Environments

In human factors and ergonomics, professionals play a key role in shaping the compatibility between tasks, products, and environments with the needs of individuals. The most successful projects demonstrate a tight integration with the human experience, underscoring the importance of ergonomic design in promoting societal well-being. A human-centered approach to design and evaluation should be championed. This book addresses this issue with an approach that focuses on environmental conditions and ergonomics in physical settings.

This book examines human factors and ergonomics across various industrial settings. From tackling challenges posed by thermal to extreme environments, the book offers an understanding of environmental ergonomics, resulting in better, more comfortable, and safer workspaces. It investigates critical topics such as G environments and +Gz Acceleration, equipping the reader with the knowledge to navigate and optimize various workplace conditions. Additionally, the reader will develop a good understanding of environmental ergonomics and understand how an appreciation of it can lead to practical and noticeable changes in their workplace.

Human Factors in Challenging Environments: From Thermal Comfort to G-Force Exposure will be of interest to human factors and ergonomics professionals seeking to deepen their expertise and broaden their understanding of the field. It will also appeal to engineers, designers, and policymakers who are invested in creating safer and more efficient work systems.

Body of Knowledge in Human Factors and Ergonomics

Series Editor: **Waldemar Karwowski**, *University of Central Florida, USA and* **Beata Mrugalska**, *Poznan University of Technology, Poznan, Poland*

This Focus book series aims to provide the perspective readers-at-large with basic knowledge of human factors and ergonomics that can be useful to students, teachers, academicians, researchers, and practitioners in domains of human activity, across all sectors of industry, business, education, home, leisure, entertainment, and sports. This new Focus series will cover the whole spectrum of topics related to a broadly defined HF/E field. All potential chapters will be selected from the content of the HF/E encyclopedia, built out and arranged into thematic short form Focus titles on the specific aspects of research, practice, and education within this field.

Should you be interested in including your own book in this series, please contact James Hobbs the Editor for Ergonomics/Human Factors and Occupational Health and Safety on james.hobbs@tandf.co.uk.

Anthropometry: Human Body Measurements and How to Use Them
Edited by Beata Mrugalska and Waldemar Karwowski

Human Factors in Challenging Environments
Edited by Waldemar Karwowski and Beata Mrugalska

Workplace Environmental Factors
Edited by Waldemar Karwowski and Beata Mrugalska

For more information about this series, please visit: https://www.routledge.com/Body-of-Knowledge-in-Human-Factors-and-Ergonomics/book-series/CRCFBK

Human Factors in Challenging Environments

From Thermal Comfort to G-Force Exposure

Edited by

Waldemar Karwowski

Beata Mrugalska

Routledge
Taylor & Francis Group
NEW YORK AND LONDON

First edition published 2026
by CRC Press
2385 NW Executive Center Drive, Suite 320, Boca Raton FL 33431

and by CRC Press
4 Park Square, Milton Park, Abingdon, Oxon, OX14 4RN

CRC Press is an imprint of Taylor & Francis Group, LLC

ISBN: 978-1-032-84131-1 (hbk)
ISBN: 978-1-032-84326-1 (pbk)
ISBN: 978-1-003-51225-7 (ebk)

DOI: 10.1201/9781003512257

Typeset in Times
by SPi Technologies India Pvt Ltd (Straive)

Contents

About the Editors

Waldemar Karwowski is a Pegasus Professor and Chairman of the Department of Industrial Engineering and Management Systems and Executive Director of the Institute for Advanced Systems Engineering, University of Central Florida, USA. He holds an M.S. in Production Engineering and Management from the Technical University of Wroclaw, Poland, and a Ph.D. in Industrial Engineering from Texas Tech University, USA. He was awarded a D.Sc. degree in Management Science by the State Institute for Organization and Management in Industry, Poland. He is the Past President of the International Ergonomics Association (2000–2003) and the Human Factors and Ergonomics Society (2007). He served on the Committee on Human Factors/Human Systems Integration, National Research Council, and the National Academies, USA (2007–2011). He has over 500 publications focused on human performance, safety, cognitive systems engineering, human-centered design, neuro-fuzzy systems, nonlinear dynamics, neurotechnology, and neuroergonomics. He serves as the Co-Editor-in-Chief of the journal *Theoretical Issues in Ergonomics Science* (Taylor & Francis).

Beata Mrugalska is an Associate Professor, Vice-Dean for Science at the Faculty of Management Engineering (2024–2028) and Head of the Division of Applied Ergonomics (2020–2024, 2024–2028), Institute of Safety and Quality Engineering, Faculty of Management Engineering, Poznan University of Technology, Poland. She holds an M.Sc. in Management and Marketing and a Ph.D. in Machine Construction and Operation and was awarded a D.Sc. in Mechanical Engineering from Poznan University of Technology, Poland. Since 2018, she has been a board member of the Center for Registration of European Ergonomists (CREE). She is responsible for promoting the professional title of EuroErgonomist in the international scientific and industrial environment. She is also a board member of the Polish Academy of Arts and Sciences, Commission of Ergonomics, and a member of the Ergonomics Committee at the Polish Academy of Sciences, Branch Poznan. She has over 150 publications on human factors in modern organizational management concepts.

Contributors

R. Ahasan
Department of Industrial Design, Division of Work Science
University of Oulu
Oulu, Finland

W. B. Albery
Biodynamics and Acceleration Branch
Air Force Research Laboratory
Ohio, USA

G. Andreoni
Dipartimento di Bioingegneria
Politecnico di Milano
Milano, Italy

R. R. Burton
Human Effectiveness Directorate
Air Force Research Laboratory
Ohio, USA

B. Chase
Department of Engineering
University of San Diego
USA

P. A. Hancock
Department of Psychology and Institute for Simulation and Training
University of Central Florida
Orlando, FL, USA

A. Hedge
Department of Design and Environmental Analysis, NYS College of Human Ecology
Cornell University
Ithaca, NY, USA

M. Kolich
Department of Industrial and Manufacturing Systems Engineering
University of Windsor
Windsor, ON, Canada

L. Maggi
Dipartimento di Bioingegneria
Politecnico di Milano
Milano, Italy

L. Piccini
Dipartimento di Bioingegneria
Politecnico di Milano
Milano, Italy

J. Salovaara
Helsinki University of Arts and Design
Helsinki, Finland

Preface

Today, human factors and ergonomics professionals worldwide contribute to the design and evaluation of tasks, jobs, products, environments, and systems in order to make them compatible with the needs, abilities, and limitations of people (IEA 2000). These professionals, who are often certified by the professional certification bodies (e.g. Board of Certification in Professional Ergonomics in the United Sates, or the Centre for Registration of European Ergonomists—CREE in the European Union), promote the human-centered approach to work systems design, testing, and evaluation, which considers the broad set of physical, cognitive, social, organizational, environmental, and other relevant human factors. Such actions should help in the socio-economic development of the world society at large. As noted by the National Academy of Engineering in the USA, in the near future "engineering will expand toward tighter connections between technology and the human experience ..., and ergonomic design of engineered products" (NAE 2004). For example, the International Ergonomics Association works to enhance the public understanding of the meaning of ergonomics and facilitates making informed decisions about the promotional claims of "ergonomically designed" systems. One of its elements is environmental ergonomics, which refers to our physiological and behavioral reactions to the ambient environment, and the design of effective barriers that allow us to survive in inhospitable settings.

This book includes not only chapters on environmental conditions but also discusses ergonomics in the context of many physical industrial settings. Its subject areas include thermal and extreme environments, G environments and +Gz Acceleration.

REFERENCES

IEA, 2000, International Ergonomics Association: www.iea.cc

NAE, 2004, *The Engineer of 2020: Visions of Engineering in the New Century* (Washington, DC: National Academy of Engineering, National Academies Press).

Environmental Ergonomics 1

A. Hedge

1.1 INTRODUCTION

Environmental ergonomics studies our physiological and behavioral reactions to the ambient environment, and the design of effective barriers that allow us to survive in otherwise inhospitable settings. Research on physiological and behavioral reactions includes studying topics such as responses to cold and heat stress, our capacity to perform work in extreme climates, and acclimatization processes. Research also focuses on requirements for clothing and enclosures that allow us to live in extreme terrestrial climates that range from deserts to Polar regions, or that allow us to venture into extremely harsh environments, such as the ocean floor or outer space. Effective barrier design encompasses the following:

- Clothing: the primary barrier that functions as our second skin and provides for thermal comfort and protection from adverse conditions (personal protective equipment), such as extremes of cold, heat, wind, wetness, air pressure, chemical contaminants, etc., as well as providing adornment and modesty values.
- Enclosures: two types of secondary barrier function as our third skin to provide safe and comfortable conditions.

DOI: 10.1201/9781003512257-1

1. Vehicles: most types of vehicles function as a protective barrier against climate conditions. Everyday transportation vehicles (cars, buses, trains, etc.) provide internal conditions within the vehicle that are regulated for human comfort. Specialized transportation vehicles (planes, submarines, space shuttle, etc.) provide a comfortable climate that allows human survival in otherwise fatal conditions.
2. Structures: climate conditions within buildings are designed to provide us with appropriate ambient environmental conditions (thermal comfort, ventilation and indoor air quality, lighting, acoustics, etc.) in even the most inhospitable terrestrial locations. Buildings regulate daily and seasonal variations in climate, and this allows people to populate a much wider range of environmental niches, well beyond those in which the clothed body alone is capable of surviving.

1.2 AMBIENT ENVIRONMENT CONDITIONS

1.2.1 Thermal Conditions

Being warm-blooded creatures, we strive to maintain a thermal balance with our environment within relatively narrow limits (37.1°C). For survival, acceptable limits for the deep body core temperature are between 35.5°C and 39.5°C. The skin can tolerate a much wider temperature range with limits between 0.6°C (skin freezes) and 45°C (skin begins to burn). Thermal balance occurs whenever net body heat production matches the rate of heat gain or loss via physiological processes. The body generates heat through metabolism and muscle activity not related to external work, and it exchanges heat with the environment via several processes. We can lose or gain heat by:

- Radiation between the skin or clothing surface and surrounding surfaces, for example, windows, sun.
- Conductance to the surfaces by direct contact with skin or clothing, for example, sitting on a warm or cold surface.
- Convection: a form of heat conduction where the surrounding air close to the body absorbs heat. Natural convection occurs as air warmed by body heat rises. Forced convection occurs when air is

moved past the body, such as by a fan, or by loose clothing (bellows effect).
- Evaporation of water through the outer layers of skin (insensible perspiration) or from the skin surface when this is wetted by sweat (perspiration) or some other external agency affects heat exchange. At maximum sweating, evaporation accounts for 15% of heat loss. Heat is also lost through warming and wetting of air that is inhaled and then exhaled, and by excretion of feces and urine.
- Storage of heat in the body. Ideally the rate of heat storage should equal zero when the body is in heat balance i.e. heat production equals heat loss with no storage.

In practice the body rarely attains or maintains heat balance, and environmental ergonomists study the many factors that influence the effectiveness of the heat exchange processes.

1.2.2 Heat Exchange Mechanisms

Three physiological mechanisms for heat exchange allow us to adapt to variations in thermal conditions. These are:

- Vasomotor: the skin needs a minimal blood supply to keep it alive. For thermoregulation this blood flow can be increased many times. Increasing skin blood flow raises skin temperature, which increases heat transfer to the environment, and cools the body core. Decreasing skin blood flow cools the skin, reduces heat transfer to the environment, and warms the body core. Changes in skin blood flow are most marked at the extremities of limbs (hands and feet) and less marked in the trunk and head.
- Sweating: transferring heat from the core to the skin becomes increasingly difficult as skin temperature approaches the body's core temperature. In hot environments sweat evaporation cools the skin surface, allowing heat transfer from the core. Sweat rate varies from a minimum of ~650 ml/day to a daily maximum continued total sweat rate of ~1 liter/h. Over shorter periods of up to 6 h the maximum short-term sweat rate may be as much as 2.5 liter/h. Sweat is secreted by two types of gland:
  1. Apocrine: forehead, back, palms of hands, armpits - produce protein-containing sweat; and
  2. Eccrine: produce dilute salty and watery sweat from all other skin areas.

- Shivering: when skin blood flow is minimal there may be excessive heat loss from the body core by conduction through the skin. Maintenance of body core temperature requires an increase in heat production and shivering is disorganized muscular activity that has this effect, and increases heat production 300–400%.

1.2.3 Effects of Heat and Cold on the Body

Our ability to successfully thermoregulate depends on individual factors, such as age, alcohol consumption, general health, and acclimatization to thermal conditions. For a normal adult, when the body core becomes too warm, blood flows from the body core to the skin, and sensible perspiration starts, which dissipates additional body heat by sweat evaporation. Hyperthermia occurs if heating of the body continues beyond its ability to lose heat through perspiration (i.e. the upper critical temperature). Hyperthermia risks are greater in hot industries, such as steel or glass-making factories, and in hot climates. Hyperthermia can be fatal, with between 240 and 1700 deaths per year in the USA, depending on heatwave conditions.

When cold, skin vasoconstriction occurs to shunt blood to the body core, piloerection (i.e. the hairs of the skin stand up) helps trap an insulating layer of air over the skin, and shivering starts. Hypothermia occurs if body cooling continues beyond our ability to generate and conserve heat (i.e. the lower critical temperature). Hypothermia risks can occur in cold industries, such as refrigerated cold storage, arctic or antarctic settings, mountain climbing, deep sea fishing or diving, outer space, etc., or in cold climates. Hypothermia can be fatal, with 600–1000 deaths per year in the USA, depending on the severity of the winter.

1.2.4 Thermal Comfort

Thermal comfort is a psychological concept that is affected by at least six variables:

- Air temperature: temperature (°C) of the ambient air.
- Mean radiant temperature: average temperature (°C) of radiant sources, such as a fire. Radiant asymmetries (warm front, cold back; warm head, cold feet) can produce thermal discomfort.
- Air speed in meters per second past the body. Air moving over the skin cools this.

- Humidity: percent relative humidity (Percent relative humidity is the ratio of the moisture content of the air divided by the maximum moisture content of that air at the same air temperature and barometric pressure, 100.) Sweat evaporation is less effective as humidity rises.
- Activity: muscular activity generates heat.
- Clothing insulation: different clothing designs and materials have different thermal insulation properties (clo units; 1 clo 58 W/m^2). Summer attire is ~0.6 clo, and a winter ensemble is ~1.0 clo. A well-padded high-back office chair insulates the body at ~0.3 clo. Clothing insulation is reduced by wind or body movements that disrupt the air layer between the skin and the clothing, or by wetting the clothing.

1.2.5 Thermal Comfort Standards

For most indoor spaces we attempt to construct settings that provide for thermal comfort. Two thermal comfort standards apply to indoor environments.

1.2.5.1 Thermal Comfort Zone (ASHRAE [American Society for Heating, Refrigerating, and Air-conditioning Engineers] 55-1981)

Thermal comfort conditions are defined as those for which 80% of the people will not report thermal dissatisfaction.

These conditions are specified as a zone on a psychrometric chart representing various combinations of air temperature and relative humidity for particular air velocities, clothing ensembles, and seasons (winter and summer). Air temperature and mean radiant temperature are assumed to be equivalent, and activity levels are assumed to be typical for indoor settings.

1.2.5.2 Fanger's Comfort Equation (ISO 7730)

Thermal comfort is defined by the physical state of the body rather than environmental conditions (Fanger 1970; ISO 1984). A comfort equation is computed that predicts comfort temperature (C), the average thermal comfort votes for any group of people (predicted mean vote, PMV), and the percentage of people who will be dissatisfied with the thermal environment (predicted percentage dissatisfied, PPD). Several experiments show that for sedentary work and light clothing, the preferred temperature is close to the 25.6°C predicted by Fanger's equation. Preferred temperatures appear to be uninfluenced by age, race, or individual differences (Fanger 1970). Studies of sex differences in thermal comfort have produced conflicting results. Recently, computer programs and portable thermal comfort instruments based on ISO 7730 have been developed. However, in some laboratory and field research, the PMV does not always give good predictions of thermal comfort ratings (Bensel and Santee 1997; Konz 1997; McIntyre 1980).

1.2.6 Temperature and Performance

Many studies show how changes in thermal conditions can affect human performance (Konz 1997; Parsons 1998). Extremes of heat or cold can decrease work performance. The effects of heat on task performance may depend upon the cognitive complexity of the task and the level of heat exposure that occurs. Changes in thermal conditions can affect worker performance in office buildings, for example, typists with no incentives or knowledge of results produce more work at 20°C than at 24°C. Students' t-test scores generally are better with 22°C–23°C than 26°C. Cold conditions can impair manual dexterity when hand skin temperatures are between 13°C and 18°C. Cold conditions can also affect psychomotor task performance.

1.2.7 Ventilation and Indoor Air Quality

Indoor air quality focuses on the nature and concentrations of air pollutants (particles, fibers, gases, or bio-aerosols) inside enclosures that can adversely affect comfort, performance, and health. Ever since the energy crisis of the early 1970s, which resulted in reductions in the design standard for mechanical ventilation rates, there has been a growing awareness of the impact of indoor air quality on comfort, health, and performance.

1.2.7.1 ASHRAE Standard 62-1989

In the USA the ASHRAE Standard 62-1989 set minimum ventilation rates for occupant comfort. Acceptable indoor air quality as "air in which there are no known contaminants at harmful concentrations as determined by cognizant authorities and with which a substantial majority (80% or more) of the people exposed do not express dissatisfaction." Occupants of modern, sealed, air-conditioned buildings may experience symptoms of the sick building syndrome that may be caused by poor indoor air quality (O'Reilly et al. 1998).

(Sick building syndrome describes complaints of non-specific symptoms among building occupants, such as headaches, sore and irritated eyes, sore throats, and flu-like symptoms.) Indoor air pollutants can affect cognitive performance, cause eye, nose, throat, and lower respiratory tract irritation, elicit acute and chronic health symptoms, provoke allergic reactions and fatal illnesses. Many factors affect indoor air quality, including outdoor air quality, ventilation system design and performance, and emissions from building materials, technology, workers, and their activities. Indoor air quality studies typically investigate a wide array of contaminants that include:

- Combustion gases: exhaust emissions of combustion gases can include:
  1. carbon monoxide: a colorless, odorless gas. It affects psychomotor task performance and is toxic at high concentrations;
  2. carbon dioxide: a colorless, odorless gas that we exhale. High carbon dioxide levels promote drowsiness and ultimately asphyxiation;
  3. nitrogen oxides: several gases that can cause mucus membrane irritation and provoke bronchospasms; and
  4. sulfur oxides: several gases that can cause mucus membrane irritation and affect lung function.
- Ozone - a colorless, pungent gas that causes mucus membrane irritation, provokes asthmatic reactions, and has toxic effects. A major component of photochemical smog that occurs in many cities.
- Volatile organic chemicals (VOC): a wide variety of carbon chain or ring compounds. Indoors, they may be emitted from construction materials, paints, adhesives, furniture, office technology, cleaning agents, smoking, etc. Studies of building products' emissions report ~80% of indoor VOC may cause eye irritation, and ~30% are suspected carcinogens. Exposure to low-level mixtures of VOC adversely affects health and performance. Chamber studies show that exposure to a mixture of 22 VOC, typical of indoor settings, produces detectable neurotoxic symptoms within 2 h of exposure to a concentration of 25 mg/m^3 (O'Reilly et al. 1998).
- Formaldehyde: a colorless gas with a pungent odor. Toxicity occurs through contact with skin and mucous membranes (eyes, nose, throat). Some people can become sensitized to formaldehyde. Exposure can induce fatigue, memory lapse, headache, difficulty sleeping, respiratory problems (including asthma), and eventually chronic respiratory disease and possibly nasal and respiratory tract cancers.

- Radon: a colorless, odorless radioactive gas that emits radiation as it decays into its progeny through a decay chain. Radon is thought to cause some 20,000 deaths per year among non-smokers.
- Bioaerosols: In the USA, respiratory tract infections (mainly viral illnesses) account for 75 million visits to physicians, 1.25 million hospitalizations, $15 billion direct medical costs, 150 million lost workdays, and $59 billion in indirect costs (lost income, absenteeism). Three groups of microorganisms, viruses, bacteria, and fungi, are responsible for many of the bio-aerosol illnesses associated with indoor exposures. Bio-aerosol exposure can be fatal, such as those deaths from Legionnaire's disease. The relationship between the risk of respiratory tract infections and the spatial design of buildings and ventilation system performance is only poorly understood. Research shows increased risk of infection in shared offices and in air-conditioned spaces (O'Reilly et al. 1998).

1.2.8 Lighting

Visible light is electromagnetic radiation that is emitted by a source. Since ancient times people have used artificial light sources, such as oil lamps or candies, to supplement or replace daylight. Today, electric lighting provides a wide array of options for creating artificial lighting conditions 24 h a day that support diverse kinds of visual work, irrespective of natural light levels.

Environmental ergonomists studying lighting pay particular attention to six factors that determine appropriate lighting for visual tasks (Boyce 1997):

- Contrast: the difference between the luminance (brightness) of an object and the luminance of the background. It is easier to see high contrast targets than low contrast targets. Contrast is affected by several factors, including the type and amount of light sources, surface reflectance, viewing angle, and target size.
- Size: the larger an object, the easier it is to see. The size of the image on the retina is important, not the size of the object, and this is affected by viewing distance. As we bring an object into closer focus, we can see finer details.
- Time: seeing takes time. The retinal biochemical processes take time to occur and visual processing and recognition also take time, especially for unfamiliar objects. Consequently, viewing time is important. We need bright light if objects are to be briefly seen.

- Luminance: the quantity of incident light arriving from a source. This determines illuminance (i.e. the amount of light falling over a unit area) which affects the luminance of the task.
- Color: this affects both contrast and luminance factors because the eye is not equally sensitive to all wavelengths, and ~8% of men and 0.5% of women have color vision deficiency (Boyce 1981).
- Visual acuity: the resolving power of the eye is affected by anatomical factors. The shape of the cornea produces ~70% of its refracting (light bending) capabilities, with the crystalline lens providing ~30%. Changes in lens curvature (accommodation) alter refraction and focus light onto the photoreceptive retina.

1.2.9 Visual Processes

Light is detected by photoreceptors in the retina of each eye. Each retina has ~130 million rods (monochromatic vision), and ~7 million cones (color vision) concentrated in the fovea, the region of greatest sensitivity and acuity. Plentiful light allows high visual acuity and color vision (photopic vision). Low light levels reduce visual acuity and impair our ability to see colors (scotopic vision). The eyes adapt slowly when moving between light and dark spaces, and visual recovery can take up to 1 h. The eyes adapt more rapidly when moving between dark and light spaces, but immediate visual function can be impaired for a short while. Partial adaptation can occur if visual field contains adjacent dark and bright areas. This is why ergonomists do not recommend placing a VDU screen either facing or backing against a bright background, such as a window.

1.2.9.1 Aging on Visual Performance

With age, especially at 40 years of age, the lens loses flexibility and presbyopia occurs. The lens also becomes increasingly opaque with age, which reduces light transmission and can result in a cataract. Fluid-filled compartments of the eye, the aqueous humor and vitreous humor, function to provide nutrients to the non-vascular structures within the eye and maintain the shape of the eye. With age, more light scattering occurs, especially at the shorter wavelengths. Retinal sensitivity decreases with age, especially at low luminance levels. Sensitivity to glare increases with age (Anshel 1990).

1.2.9.2 Lighting, computer work, and eyestrain

Workers, especially computer users, may experience glare in their workplace. Disability glare is direct glare from a bright light source, such as bright sunlight through a window Discomfort glare is indirect glare from uncomfortably bright reflections, such as ceiling light reflections in a computer screen. Screen reflections can be distracting and cause visual discomfort and eyestrain. Surveys of computer users find that inadequate lighting is frequently reported, and eyestrain afflicts almost half of all workers. Field studies testing different lighting have found that computer workers report fewer complaints of eyestrain and eye focusing problems, greater satisfaction, and express strong preferences for lensed-indirect uplighting compared with parabolic lensed downlighting (Hedge et al. 1995).

1.2.10 Acoustic Environment

Sounds are an integral part of human communication for most people. Sound is the result of pressure changes in the air, and its intensity is measured in decibels (dB). A very quiet room would have a sound level at ~40 dB, conversations are mostly between 55 and 65 dB, and a jet taking off may emit sound 130 dB. Sounds 120 dB usually feel painful to the ears. Brief exposure to high-intensity sounds can produce a transient impairment in hearing, called a temporary threshold shift. Prolonged exposure to loud noise can cause a permanent threshold shift and permanent hearing damage.

When sounds are intense, annoying, or otherwise unwanted they become noise and that can adversely affect our performance and health. The noisiness of a sound often results from the subjective impression of how annoying it is. Noise can be unwanted sound that carries unpleasant information about the sound source (e.g. a crying baby), or it can be loud, annoying unwanted sound (e.g. a jackhammer outside one's window late at night). The ear is least sensitive to lower sound frequencies, and most sensitive to frequencies in the speech range, especially ~3 kHz. Lower frequency sounds are generally less annoying than higher frequency sounds of the same subjective loudness (Kryter 1985).

1.2.10.1 Occupational Noise

Sources of occupational noise include the vibrations of structures, machines, or their components, and also aerodynamic turbulence (e.g. high-pressure air jet). Occupational noise exposure is assessed by quantifying a permissible noise dose that is based on daily exposures. Permissible noise dose is regulated by OSHA (Sanders and McCormick 1993). Industrial plants typically transmit noise to the external environment via certain sources or routes (e.g. open windows, roof ventilators, steam injectors, compressors, diesel engines). Planning regulations have reduced the impact of noise-generating industries on many residential areas. However, indoors in manufacturing industries, 80% of noise levels are around 80 dB, and 20% reach 95 dB. Exposure to noise this loud can cause hearing loss (i.e. physiological deterioration of hearing due to destruction of hair cells in the organ of Corti, and decreased number of associated nerve fibers). Exposure to uncontrollable and unpredictable noises can also have stressful aftereffects on behavior. Studies of occupational noise show that this is associated with annoyance, health problems, accidents, and reduced efficiency of performance (Crocker 1997).

1.2.10.2 Speech Privacy

In many modern open-plan offices, worker performance is impaired by inadequate speech privacy and distracting noise from computers, telephones, printers, photocopiers, air conditioning, and conversations. Speech privacy increases as the intelligibility of encroaching sounds decreases. Solutions to office noise problems include increasing the sound absorptance of room surfaces, especially the ceiling, and reducing source strength (e.g. choosing quieter equipment, encouraging workers to speak more quietly). Sound masking systems can also be effective because they add background "white" noise to the office that lowers the signal-to-noise ratio.

1.2.11 Electromagnetic Fields

Electric and magnetic fields (EMF) emanate from power lines and from office equipment, such as cathode ray tube computer screens and system units with power transformers. There are concerns about prolonged EMF exposures and health effects, such as nervous system cancers and leukemia. Research findings are conflicting. Most modern computer monitors are low-emission designs, and with older computers, EMF issues can be minimized by putting the computer monitor and system unit at arm's length on a non-conductive surface.

1.2.12 Vibration and Motion

Oscillating motions can cause vibrations that can adversely affect the body and performance in several ways (Griffen 1997). Whole-body vibration occurs when the body is supported by an oscillating surface, such as a vibrating truck seat or a boat moving up and down on water. Very low frequency whole-body oscillations (usually 1 Hz) can cause motion sickness. Whole-body vibration also reduces efficiency for manual tasks such as tracking, typing, or writing. Segmental vibration occurs when only a part of the body is in contact with an oscillating surface. Sometimes this is useful, for example, a vibrating pager silently signals a call.

Often the segmental vibration is associated with hand-transmission from a tool, such as a chainsaw, weed trimmer, or jackhammer. In extreme cases hand-transmitted vibration can lead to a disease called vibration white finger.

1.2.13 Gravity

Riding a roller coaster subjects the body to high rates of motion and brief fluctuations in gravity. For some people this is a thrill, for others a fright, and for some a cause of motion sickness. These transient changes are magnified for people in other high-speed settings, for example, a pilot flying a high-G turn, an astronaut during lift-off. High-G situations affect the body in many ways, increasing loads on the musculoskeletal system and modifying blood distribution, which sometimes causes symptoms ranging from elbow pain to loss of consciousness. At the other gravitational extreme our continued exploration of long-term effects of space requires knowledge about the effects of microgravity on the body and on performance (Albery and Woolford 1997).

1.3 CONCLUSIONS: WORK–ENVIRONMENT DESIGN: A PROCESS OF INTEGRATION

Environmental ergonomics is making an increasingly important contribution to improving the design of environments. Poorly designed workplaces can create many problems for occupants that affect their attitudes and motivation (e.g. satisfaction with work), their health and stress level, and reduce their efficiency. Well-designed work environments provide good ambient conditions that satisfy the physiological and personal requirements of workers. Environmental ergonomists typically use a systems approach to analyze the interactions between users, technology, the organization, and the ambient environment. They also work as members of a design team that includes other professionals, such as architects, interior designers, and engineers (ventilation, illumination, acoustics). In many respects environmental ergonomics is still in its infancy, yet as human influence spreads into remote and hazardous places on and beyond the earth, environmental ergonomics has an increasingly important role to play.

REFERENCES

Albery, W. B., & Woolford, B. (1997). Design for microgravity and microgravity environments. In G. Salvendy (Ed.), *Handbook of human factors and ergonomics* (2nd ed., pp. 935–963). John Wiley & Sons.

Anshel, J. (1990). *Healthy eyes better vision*. Body.

ASHRAE (1981). *Standard 55-1981: Thermal environmental conditions for human occupancy*. American Society of Heating, Air Conditioning, and Refrigerating Engineers.

ASHRAE (1989). *Standard 62-1989: Standard for ventilation for acceptable indoor air quality*. American Society of Heating, Air Conditioning, and Refrigerating Engineers.

Bensel, C. K., & Santee, W. R. (1997). Climate and clothing. In G. Salvendy (Ed.), *Handbook of human factors and ergonomics* (2nd ed., pp. 909–934). John Wiley & Sons.

Boyce, P. R. (1981). *Human factors in lighting*. Applied Science.

Boyce, P. R. (1997). Illumination. In G. Salvendy (Ed.), *Handbook of human factors and ergonomics* (2nd ed., pp. 858–890). John Wiley & Sons.

Crocker, M. J. (1997). Noise. In G. Salvendy (Ed.), *Handbook of human factors and ergonomics* (2nd ed., pp. 790–827). John Wiley & Sons.

Fanger, P. O. (1970). *Thermal comfort*. McGraw-Hill.

Griffen, M. (1997). Vibration and motion. In G. Salvendy (Ed.), *Handbook of human factors and ergonomics* (2nd ed., pp. 828–857). John Wiley & Sons.

Hedge, A., Sims, W. R., & Becker, F. D. (1995). The effects of lensed-indirect uplighting and parabolic downlighting on the satisfaction and visual health of office workers. *Ergonomics, 38*(2), 260–280. https://doi.org/10.1080/00140139508925103

International Standards Organization. (1984). *International Standard 7730: Moderate thermal environments-Determination of PMV and PPD indices and specification of the conditions for thermal comfort*. ISO.

Konz, S. (1997). Toxicology and thermal comfort. In G. Salvendy (Ed.), *Handbook of human factors and ergonomics* (2nd ed., pp. 891–908). John Wiley & Sons.

Kryter, K. D. (1985). *The effects of noise on man*. Academic Press.

McIntyre, D. A. (1980). *Indoor climate*. Applied Science.

O'Reilly, J. T., Ragan, P., Gots, R., & Hedge, A. (1998). *Keeping buildings healthy: How to monitor and prevent indoor environmental problems*. John Wiley & Sons.

Parsons, K. (1998). *Human thermal environments*. Taylor & Francis Group.

Sanders, M. S., & McCormick, E. J. (1993). *Human factors in engineering and design* (7th ed., pp. 589–681). McGraw-Hill.

FURTHER READING

ANSI/ASHRAE Standard 55-2024, *Thermal Environmental Conditions for Human Occupancy* (ANSI Approved).

ASHRAE Standard 62.1-2023, Ventilation and Acceptable Indoor Air Quality (ANSI Approved).

ISO 7730:2005 Ergonomics of the thermal environment — Analytical determination and interpretation of thermal comfort using calculation of the PMV and PPD indices and local thermal comfort criteria.

2 Thermal Comfort

P. A. Hancock

2.1 PREAMBLE

Trying to provide a small, concise essay on thermal comfort, defined as "those conditions of mind that express satisfaction with the thermal environment," is a difficult task because it requires the integration of ideas from areas as diverse as thermal physics, physiological systems, environmental engineering, and human perception. Thus, a commentator faces the pervasive problem of breadth versus depth of coverage. I try to provide a compromise between these two and where space restricts a detailed exposition, I can fortunately refer the reader to most useful in-depth treatments such as those presented by Chatonnet and Cabanac (1965), Fanger (1967, 1970), Parsons (1993), Bensel and Santee (1995), and most recently (Konz 1995, 1997).

2.2 INTRODUCTION

Perhaps the most appropriate place to begin an essay on thermal comfort is within the realm of physiology and especially with the recognition of the regulatory nature of many human physiological processes. Following upon Harvey's discovery of the circulation of blood, perhaps the most important observation on physiological systems was Claude Bernard's identification of the "fixité de la milieu interieur"—the "stability of the internal environment." It is clear that

DOI: 10.1201/9781003512257-2

all organisms have, to some extent, to provide such stability in order to deal with the vagaries of an uncertain external environment. This principle holds across multiple levels of detail, ranging from the cellular level to the level of the organism itself. Establishment of a stable biochemical platform upon which life processes can proceed without the momentary interference of the "outside" world is crucial. As a consequence, living systems are not "closed" with respect to the environment (they do interact with the environment), but neither are they completely "open" (in which case they would slavishly follow environmental variation). Therefore, living systems exert hybrid control, opposing environmental influences on occasion while selectively permitting a degree of influence at different times (Kenshalo *et al.* 1961; Benzinger 1963).

In his original observations, Claude Bernard referred to temperature control as a primary example of regulation, and regulatory control was also a principal concern of Cannon (1932) in his classic text, for which he employed the term "homeostasis." While the interchange between the human and the environment is more free and dynamic than the term homeostasis implies, the fundamental idea of a temperature balance as a crucial foundation for existence persists. Like many other organisms, human beings are homeotherms. That is, they maintain a resting body temperature that is well above the temperature of the environment that surrounds them. This characteristic is advantageous in that humans can explore many places that would not support the existence of organisms that rely principally upon the environment to warm them. The penalty for such freedom, however, is a constant demand for energy to support this elevated temperature. Thus, many animals, and until relatively recently human beings, had to spend much of their time in search of food. What this means is that in most terrestrial conditions, human beings are losing heat to the environment, and the rate of this loss is governed by physical laws of heat exchange. Of course, there are hot deserts where this tendency is reversed and the unaided individual experiences considerable heat stress. In order to buffer the effects of excessive heat loss or heat gain, human beings have adopted various forms of clothing and covering. While today clothing serves many different functions, the primary impetus for clothing was as a tool or aid in the process of thermal comfort regulation.

2.3 DESCRIBING THERMAL INTERACTION

Body temperature is a crucial facet of existence and a vital mediator of many critical biochemical reactions. Activity and performance follow body temperature such that performance is poor when temperature drops in the small hours of the morning and is optimized in the late afternoon and early evening as the circadian rhythm peaks. Disorders of temperature are often symptomatic of more widespread disturbance or illness, and excessively high or low body temperatures themselves can result in illness and eventually death. Thus the successful control of body temperature, disturbances to it, and failures of regulation are crucial concerns of the medical, environmental, and behavioral sciences (Hardy 1971). To begin to understand thermal regulation, it is necessary to describe the avenues of thermal interaction between the human and the environment. One of the first such descriptions was given in the form of an equation by Burton (1934):

$$M + W - E \pm R \pm C \pm S = 0 \tag{2.1}$$

where M is the basal metabolic rate, W is the work rate the individual is engaged in, E is the rate of evaporation of moisture from the body, R is the radiant heat lost or gained from the environment, C is similarly the convective heat lost or gained from the environment, and S is the storage of loss of heat expressed as a change in the deep body temperature of the individual involved.

The equation is stated such that the outcome is zero, which is perhaps the easiest way to see that Burton's model follows the fundamental idea of homeostasis as a highly stable (i.e., zero change) state. As has been noted, this is too rigid a conception and where such rigidity must be tempered provides important clues to the achievement and sustenance of thermal comfort. First, Burton's equation treats the human being in exactly the same fashion as it would treat an inanimate object. What is missing from his picture is the idea of human goal-directed behavior. The idea of behavioral regulation is critical as individuals engage constantly and primarily in behavioral regulation, in part to obviate the need for more effortful physiological regulation. Consider for a moment a specific example. The most effective form of physiological regulation against excessive temperature is sweating. However, on a hot day, the first thing we do in reaction to the weather is not immediately to start sweating. We may stay inside air-conditioned buildings, or if we are forced outside, we might choose shaded areas to sit in. To reduce thermal stress, we might seek a cooling drink or use a fan to increase evaporation (much of which comes initially from loss from the respiratory tract). Each of these behavioral strategies *precedes* sweating as a way of maintaining temperature balance. Indeed, if these strategies are successful, we might never get to the stage of sweating at all. Thus, while simple linear equations describe heat exchange, they do not capture the subtlety and nuances of behavioral regulation, the prime goal of which is the maintenance of comfort. Hancock and Warm (1989) argued that the primary symptom that describes the limits to physiological compensation is disturbance to the dynamic steady state, for example, an uncompensated increase in body temperature. In a directly comparable way, the onset of discomfort is the symptom of the limit to behavioral regulation. As a result of this definition and this recognition, we can provide a comfort equation in the form:

$$C = f(\mathrm{Br}) \pm f(\mathrm{Pr}) \tag{2.2}$$

where C equals the perceived level of comfort, which is expressed as a function (f) of behavioral regulation (Br) combined with a function (f) of the physiological regulation (Pr). Comfort can then be expressed on a 0–100 scale, in which 100 represents a nominal optimal comfort and 0 represents loss of consciousness as a result of the intensity of the stress. Within the present equation, we can describe some points that indicate the relationship indicated is more than a descriptive one. For example, assuming that C varies between 0 and 100, where 100 represents optimal comfort, and that Br and Pr also vary between 100% effective and 0% effective, then the value of C is 0 when the value of Br is 0 and the value of Pr is 0. In principle, this general equation can apply to any aspect of comfort, which is a multidimensional construct. However, for the present purposes, we can ask what this means in terms of thermal comfort.

2.4 ASSESSMENT OF THERMAL COMFORT

To assess thermal comfort, a fairly standard experimental procedure has often been employed. An individual is exposed to a thermal condition whose physical characteristics are predefined. Typically, these factors include air temperature, relative humidity, air velocity, and radiant heat. In some of the first such experiments, individuals were asked to walk between two rooms and compare their warmth. By matching different combinations of dry bulb temperature, wet bulb temperature and air velocity (derivations of the first three factors noted above), Houghton and Yagloglou (1923) were able to construct an effective temperature (ET) scale (see Parsons 1993, Figures 7.1 and 7.2, pp. 134–135) that could be used to assess thermal comfort. For a number of reasons, the ET scale is not widely used today, although the idea of exposing individuals to different environments and asking them to comment on their comfort rating is still used (Fobelets and Gagge 1988). For example, Rohles and Nevins (1971) derived a relationship between thermal sensation and wet and dry bulb temperatures from just such a set of procedures. As Parsons (1993) correctly indicates, it is important to ascertain whether such measures are made immediately after exposure to the new condition, since such a procedure emphasizes transient influences (see also Houghton and Yagloglou 1923), or whether requests for the perception of thermal comfort are given after the individual has had a chance to adjust to the new conditions for some extended period (see Nevins *et al.* 1966).

Fanger (1970) elaborated upon this conception through exploitation of Burton's (1934) heat balance equation described previously. Fanger realized how crucial fluid exchange was to thermoregulation, especially given that in conditions close to comfort, much of the fluid loss was imperceptible to the individual, through such avenues as respiration. Fanger was also careful to introduce physical heat transfer modifications as a function of clothing. Given that an individual was in a state of thermal balance and that sweat rate and skin temperature were within comfortable limits, Fanger developed his comfort equation. The American Society for Heating, Refrigeration and Air Conditioning Engineer's standard comfort zone (see Figure 2.1) is derived

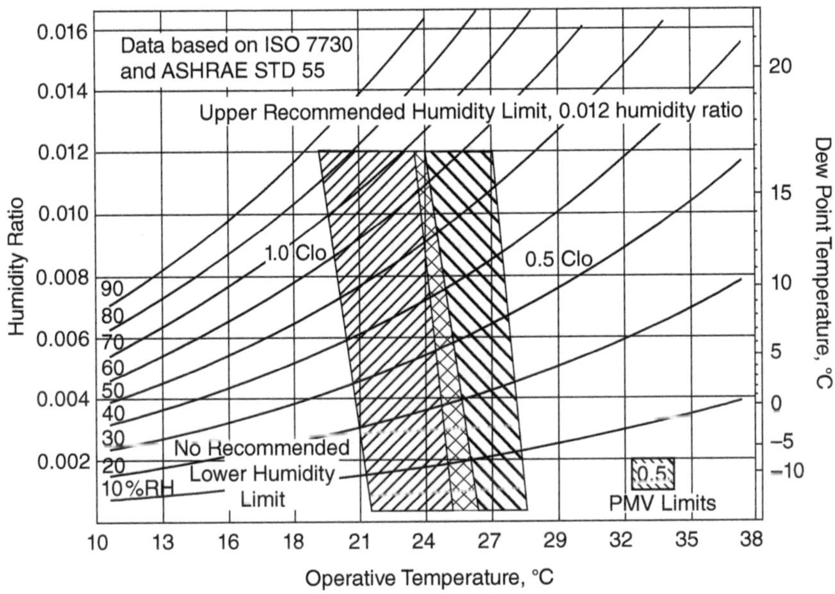

FIGURE 2.1 The ASHRAE comfort zone shown on a psychometric chart. The slanting dashed lines have equivalent skin wetness and thus approximately the same comfort. Thus, a dry bulb temperature (DBT) of 25°C and 80% relative humidity has the same comfort as a DBT of 27.5°C and 20% relative humidity. The line is given a label of "new effective temperature" or ET* where the dashed line crosses the 50% relative humidity line; in this example, it is 26.1°C ET*.

(Reproduced from Konz 1995).

from this approach using dry bulb temperature, water vapor pressure, air velocity, radiant temperature, metabolic rate, clothing insulation, and exposure time. Comfort votes were used to derive acceptable levels, and 94% of individuals described themselves as "comfortable," "slightly warm," or "slightly cool" within the illustrated zone. The actual thermal comfort vote could be predicted using the following equations:

$$\begin{aligned} \text{TS} &= -1.047 + 0.158\,\text{ET} && \text{ET}^* < 20.7 \\ \text{TS} &= -4.444 + 0.326\,\text{ET}^* && 20.7 < \text{ET}^* < 31.7 \\ \text{TS} &= 2.547 + 0.106\,\text{ET}^* && \text{ET}^* > 31.7 \end{aligned} \tag{2.3}$$

where TS is the thermal sensation vote for individuals engaged in sedentary activity at 0.6 clo, clothing insulation value.

As Konz (1995) notes, even within these boundaries, 3% of individuals will vote that they are "cool" and another 3% that they are "hot," and thus comfort is a population-distributed parameter and not an exact value. There are also variations for winter and summer conditions, largely because of clothing preferences and further details of this can be found in Konz (1995).

These indices have been used by different scientific, advisory, and regulatory agencies (cf., ASHRAE 1989; ISO 1992) as recommendations to professionals such as architects and civil engineers for use in their work. What is clear is that the major efforts to understand human thermal comfort have largely focused on derivations from heat transfer physics, and to date, relatively little attention has been paid to behavioral adaptation beyond the use of comparative psychophysical approaches.

For example, the equation presented by Fanger and employed by a number of agencies relies almost completely upon physical values with a subsequent reference to a vote of dissatisfaction. The crucial question becomes, what do dissatisfied people do about their (thermal) environment? The implication of the physical standards is that they continue their present activity at its present level and go through differing degrees of unhappiness over differing specified periods of time. But, of course, this does not happen. Individuals modify their thermal environment almost continuously. They add or shed clothing, take cooling or warming drinks, change and alter the location of furniture, and add a small fan or small air-heater to their immediate locale. In fact they engage in a myriad of behavioral adjustments, many of which far supersede in influencing the variation of one or two degrees of temperature or percentage points of humidity. In practical terms, the challenge of the environmental engineer is to present an envelope of acceptable conditions within which the individual is free to engage their own adjustment strategies, as is beginning to be seen in production automobiles. The challenge is to maintain an acceptable "envelope" of conditions while outside a permanent facility. In really hostile conditions, such as the deep desert, the deep oceans, and deep space, this represents a significant problem.

2.5 SUMMARY

Thermal comfort is an important issue. Individuals expected to perform outside the ranges of thermal comfort for extended periods will be ineffective, unsafe, unwell, and eventually injured. While forms of forced acclimatization can help expand the range of tolerance, this strategy is an attempt to "fit the human to the conditions" in contrast with the efforts of human factors and ergonomics that endeavor to "fit the conditions to the human." Present indices are founded upon engineering approaches that deal with the occupant largely as though they were an insensate object. However, we know this is not true and human beings act to behaviorally regulate their environment as much as they act to physiologically regulate functions of the body. Designers need therefore to facilitate this self-directed, goal-related activity and, as much as possible, establish an "envelope" of acceptable conditions with sufficient user freedom to easily manipulate local micro-environments. Such opportunities are now becoming available as options on advanced vehicles. Further effort is clearly required to clarify the effects of individual differences on thermal comfort that do not seem related to different intrinsic appreciation of warmth but rather due to the efficiency of behavioral regulation in modulating local conditions on a regular basis. Finally, the extensive database on thermal comfort can be used as a foundation for further research into the larger issue of the more general concept of comfort itself and how it relates to human performance and capability in general.

ACKNOWLEDGMENT

I am especially grateful for the comments and insights provided by Professor Stephan Konz on an earlier version of this chapter, which were especially helpful in the revision of the work. Any remaining errors are those of the author.

REFERENCES

American Society of Heating, Radiation, and Air-Conditioning Engineers (ASHRAE). (1989). *Thermal environmental conditions for human occupancy*. ANSI/ ASHRAE.

Bensel, C. K., & Santee, W. R. (1995). Climate and clothing. In G. Salvendy (Ed.), *Handbook of human factors* (pp. 912–928). Wiley.

Benzinger, T. H. (1963). Peripheral cold and central warm reception: Main origins of human thermal discomfort. *Proceedings of the National Academy of Sciences, 49*(6), 832–839.

Burton, A. C. (1934). The application of the theory of heat flow to the study of energy metabolism. *Journal of Nutrition, 7*, 497–533.

Cannon, W. B. (1932). *The wisdom of the body*. W. W. Norton.

Chatonnet, J., & Cabanac, M. (1965). The perception of thermal comfort. *International Journal of Biometeorology, 9*(2), 183–193.

Fanger, P. O. (1967). Calculation of thermal comfort: Introduction of a basic comfort equation. *ASHRAE Transactions, 73*, 4.1–4.18.

Fanger, P. O. (1970). *Thermal comfort*. Danish Technical Press.

Fobelets, A. P. R., & Gagge, A. P. (1988). Rationalization of the effective temperature, ET*, as a measure of the enthalpy of the human indoor environment. *ASHRAE Transactions, 94*(1), 12–31.

Hancock, P. A., & Warm, J. S. (1989). A dynamic model of stress and sustained attention. *Human Factors, 31*(5), 519–537.

Hardy, J. D. (1971). Thermal comfort and health. *ASHRAE Journal, 77*, 43.

Houghton, F. C., & Yagloglou, C. P. (1923). Determining equal comfort lines. *Journal of the American Society of Heating and Ventilation Engineers, 29*, 165–176.

International Standards Organization (ISO). (1992). *DIS 7730: Moderate thermal environments—Determination of the predicted mean vote (PMV) and predicted percentage dissatisfied (PPD) indices and specification of the conditions for thermal comfort*. ISO.

Kenshalo, D. R., Nafe, J. P., & Brooks, B. (1961). Variations in thermal sensitivity. *Science, 134*(3473), 104–105.

Konz, S. A. (1995). *Work design: Industrial ergonomics*. Holcomb Hathaway.

Konz, S. A. (1997). Toxicology and thermal comfort. In G. Salvendy (Ed.), *Handbook of human factors*. Wiley.

Nevins, R. G., Rohles, F. H., Springer, W., & Feyerherm, A. M. (1966). A temperature humidity chart for thermal comfort of seated persons. *ASHRAE Transactions, 72*, 283–291.

Parsons, K. C. (1993). *Human thermal environments*. Taylor & Francis Group.

Rohles, F. H., & Nevins, R. G. (1971). The nature of thermal comfort for sedentary man. *ASHRAE Transactions, 77*, 239–246.

FURTHER READING

Ávila, S., Mrugalska, B., Wyrwicka, M. K., Souza, M., Ávila, J., Cayres, É., & Ávila, J. (2020). Cognitive and organizational criteria for workstation design. In Karwowski, W., Trzcielinski, S., Mrugalska, B. (Eds.) *Advances in Manufacturing, Production Management and Process Control. AHFE 2019. Advances in Intelligent Systems and Computing*, vol. *971*. Springer. https://doi.org/10.1007/978-3-030-20494-5_15

Górny, A. (2017). The use of working environment factors as criteria in assessing the capacity to carry out processes. *MATEC web of conferences* (Vol. *94*, p. 06011). EDP Sciences.

Mansi, S. A., Barone, G., Forzano, C., Pigliautile, I., Ferrara, M., Pisello, A. L., & Arnesano, M. (2021). Measuring human physiological indices for thermal comfort assessment through wearable devices: A review. *Measurement*, *183*, 109872.

Su, X., Yuan, Y., Wang, Z., Liu, W., Lan, L., & Lian, Z. (2024). Human thermal comfort in non-uniform thermal environments: A review. *Energy and Built Environment*, *5*(6), 853–862.

Zheng, P., Wu, H., Liu, Y., Ding, Y., & Yang, L. (2022). Thermal comfort in temporary buildings: A review. *Building and Environment*, *221*, 109262.

3 Active Customized Control of Thermal Comfort

G. Andreoni, L. Piccini, and L. Maggi

3.1 INTRODUCTION

Thermal comfort provision is a relevant aspect of the project of artificial environments. The quality of life in homes, offices, and vehicles is strongly dependent on the capacity to reach and maintain the best ambient condition for the occupant. Today, the possibility of realizing personalized HVAC (heating, ventilation, and air-conditioning) systems and providing a more distributed and personalized climatic control seems no longer illusionary, however, despite this fascinating and interesting perspective, the control of these and classical systems is not so efficient. It is worth noting that, in our daily life experience, obtaining a fully comfortable environment may be extremely difficult; for instance, how often do we discuss fan temperature and ventilation set in our work place? In fact, to date, there is a great gap between the potentiality offered by HVAC technologies, and the way of controlling those systems to satisfy the user's comfort.

The physical phenomena are well described with the energy balance of the body, an accurate physical description including both the biological and the environmental phenomena; however, from this model it is impossible to extract all the necessary information to obtain indications regarding the well-being of

DOI: 10.1201/9781003512257-3

the subjects. In fact, thermal comfort is a mind condition that is related to a psychophysiological state that is impossible to describe mathematically from a biological point of view, but it can be observed and described from a social point of view using mathematical models extracted with statistical analysis of experiments conducted on a large number of people.

In 1967, Fanger proposed (Fanger 1967; Charles 2003) a general and complete indicator of personal comfort for the inner environment and proposed a mathematical representation of this condition, attempting to create an indicator to quantify the level of comfort and the percentage of people who consider this standard really comfortable. Other researchers proposed different models closer to the physical one, but the one proposed by Fanger is the most diffused and has been acknowledged in the specifications for the project and the design of buildings (ISO: 7730, 1997). In this context, it is conceivable to gather all the information and data required for the Fanger model, to attempt to estimate the comfort level of the ambient to realize a feedback system to control the inner climatic condition. In our work we have stated the possibility of realizing an integrated wearable and environmental system to establish the personal thermal comfort, using a set of sensors and transducers that measure temperature and relative humidity in the room and on the people living or working there.

3.2 THE AUTOMATIC ANALYSIS OF THERMAL COMFORT

3.2.1 The Thermal Comfort

The design of automatic control of thermal comfort presents two main problems:

1 What are the physical parameters involved in the description of the phenomena?
2 What is the relationship between them and the psychophysiologic condition?

From the biological point of view, our autonomic nervous system (ANS) tries to continuously maintain the homeostatic condition using the set point generated in the hypothalamus. From the physical point of view, mammals' thermal equilibrium can be described with an energetic balance that involves many different factors, related to the biological functionality of the human body and the interaction with the surrounding environment. The difference between the energy we produce, the energy used to fulfill biological functions, and the energy lost in the form of heat sent to the environment can lead our body out of the preferred thermal condition. The equilibrium is described in more detail in the equation shown below which is normalized on body surface area:

$$M - L - U - E - V = R + C\left(\mathrm{W} / \mathrm{m}^2\right) \tag{3.1}$$

where M is the metabolic energy produced; W is the external work; U is the amount of energy stored in the organism; E is the heat lost in perspiration; V is the heat lost in ventilation phenomena; C is the heat lost in convection phenomena; and R is the energy lost for the irradiation phenomena.

This formula provides us with the information concerning the most important contributions that we have to consider to establish whether the thermal equilibrium is reachable or not, but it does not relate this condition to the psychophysiological state. In order to overcome this problem, Fanger modified this model, identifying a relationship between M, R, the clothes, the environmental temperatures, and the human well-being for a wide number of subjects. This model can be suitable for inner environments with moderate climate change but represents a large number of working activities, domestic, ambient, and vehicles, dealings with many scenarios.

3.2.2 Fanger's Model of Thermal Comfort

Fanger identified a model in which the main factors involved in the heat exchange are strictly dependent only on environmental conditions (R, C) and on an insulating barrier between the body surface and the air. This model was accepted by the International Standard Organization and introduced in the ISO 7730 norm for the comfort inside moderate thermal environments. Fanger created an index that expresses the satisfaction of people in well-defined situations: the predicted mean vote (PMV) that describes, with the seven different discrete levels of the ASHRAE comfort scale, the feelings of the considered subjects (e.g. −3 cold, −2 slightly cool, 3 hot).

We would maintain that there is no necessity to specify comfort, and thereby PMV, with a more precise scale, because considering the psychophysics connotation, a thermal comfort index has to be intrinsically at low resolution. The second parameter proposed, the PPD (percentage of dissatisfied), is a function of PMV and indicates the number of people who are not in a comfort condition at that PMV, thereby it completes the scenario proposed for an automatic and personalized thermal comfort, forcing the system to also obtain a specific PPD (typically 10% of the population).

In more detail, Fanger proposed to quantify the thermal comfort using Equation 3.2:

$$\begin{aligned} PMV = & \left(0.303 * e^{-0.036M} + 0.028\right) * \{(M - W) \\ & - 3.05 * 10^{-3} * \left[5733 - 6.99(M - W) - p_a\right] \\ & - 0.42 * \left[(M - W) - 58.15\right] - 1.7 * 10^{-5} M (5867 - p_a) \\ & - 1.4 * 10^{-3} M * (34 - t_a) - 3.96 * 10^{-8} f_{cl} * [(t_{cl} + 273)^4 \\ & - (t_r + 273)^4] - f_{cl} * h_c (t_{cl} - t_a)\} \end{aligned} \tag{3.2}$$

where M is the metabolic energy produced (*met*, 1 met = 58.2 W/m^2); W is the free energy production (external work) (W/m^2); p_a is the partial vapor pressure of water vapor (P_a); t_a is the air temperature (°C); f_{cl} is the ratio between the covered and free surface of the body; h_c is the convective heat exchange coefficients (W/m^2/°C); t_{cl} is the clothes temperature of the external surface (°C); and t_r is the mean radiant temperature (°C).

It is worth noting that this formula is composed of many terms that are not easy to calculate, such as M, W, t_{cl}, h_c, and t_r, and we consequently have to consider the possibility of extrapolating these terms in an automatic way. First of all, the metabolic energy depends on the subject's activity, therefore, the three possible methods of obtaining this information are: asking subjects, pre-setting the values for the ambient which has to be controlled, or monitoring subjects. The third solution is extremely complex and expensive, whereas the first two are more practical and require very limited human intervention; thereby it is conceivable to assume that the metabolic activity slowly and modestly changes as, for instance, during daily work in an office. M values are reported in the ISO documents, for instance, a sedentary work produces nearly 70 W/m^2, 1.2 met, while a moderate standing activity needs about 116 W/m^2, 2 met.

Assuming W is negligible in many activities, the crucial topic is how to consider the environmental contribution since it depends on different temperatures and other physical parameters that are involved in the process of heat transmission. There are many instruments that can measure all the required physical parameters except the clothes influence, but such systems only make a global measure available that does not allow a personalized thermal comfort system to be realized; moreover, they are not cost-effective and do not suit typical working scenarios. In this analysis, we explain the assessment of the possibility of using Fanger's model, providing all the information it requires, exploiting solutions of ambient intelligence and domotics systems.

As proposed previously, all the required variables can be obtained in different ways, with real-time monitoring or presetting the values when possible, as presented below. The steam pressure, p_a, is a well-known function of air temperature and relative humidity (RH), which can be measured with appropriate transducers while h_c depends on t_{cl}, t_a, and v_{ar}; the air velocity relative to the human body [m/s] depends on the HVAC system (without doubt, today, it is the main source of airflow in an inner environment) and activities that are carried out in the considered place: for instance it is possible to directly control the state of fan coils, to preset the level of subjects movement, or preset the global air velocity using the values reported in the standards.

Although it is not possible to preset the remaining parameters (surface temperature of the clothes, the covered surface ratio, and the mean radiant temperature) because they are strongly dependent on local climate and subjects position, the new technologies in the field of wearable computing and environmental sensors can help in the measure of these variables, avoiding expensive and unsuitable systems in relation to the aims reported.

3.2.3 Introduction of New Technologies to Assess Comfort

In order to explain the possible solutions, it is important to emphasize the new capabilities of commercial devices compliant with ambient intelligence applications. Today, it is possible to obtain a smart measure of many physical parameters using a large and wide use of miniaturized sensors and transducers that can be placed in the environment, on clothes and inside accessories providing measures in a transparent way, without interfering with the subjects usual daily activities: this is the new concept of unobtrusive (non-intrusive) measure (Di Rienzo *et al.* 2004). These sensors, with low-power radio frequency (RF) transmitter systems, can interact with the central station that can process data to analyze them and, at the end, obtain a good approximation of the PMV value. By placing thermometers on walls, desks, windows, ceilings, chairs, or whatever would be required for the specific context, it is possible to measure the surface temperature of the parts that offer a contribution to the mean radiant temperature, usually measured with specific instruments but which requires complex computation (Meijer and Van Herwaarden 1994). Considering the relative distance from subjects to these different entities, and the view factor which is dependent on the geometry of the environment and the furniture, with a weighted sum of the different contributes, it is possible to simplify the problem and successfully approximate t_r which is one of the most important variables for the identification of the PMV value.

The contribution of t_{cl} in Fanger's model is assumed as a function of the clothes thermal resistance, measured in clo (1 clo = 0.155 m^2°C/W), that is also standardized.

It seems possible to correlate these values of clo, with the difference between the external surface temperature of clothes and the inner one, in order to realize an automatic system. In more detail, it is conceivable to realize garments with embedded devices (Di Rienzo *et al.* 2004) that are able to locally measure temperature with a distribution as large as required to correctly estimate the heat diffusion through the interface. The humidity measure is related more to the metabolic activity of the subject but can also provide indications about the kind of clothes. The realization of the new paradigm of measurement is the major breakthrough offered by the wearable system, allowing monitoring of the subject's thermal surface conditions and analyzing and understanding the clothes influence.

3.3 SYSTEM DESIGN

3.3.1 Description

The optimal system solution strongly depends on the environment size, geometry, and the position of the subjects. Each sensor could be capable of measuring temperature and relative humidity, the sensors can be controlled in different ways, for example with a microcontroller, and provide different sampling frequencies to satisfy the specifications. The sensors network presented can be easily connected to a computer where the HVAC control system resides, and the PMV can be calculated.

3.3.2 Design Specifications

In the previous chapter we argued about the problem of estimating PMV. Considering that analysis, we can understand and outline the design specification we have to target in order to build a system that satisfies the requirements that were introduced. How many measurements do we need during a certain amount of time? How accurate should those measurements be? Some design aspects directly arise from the technical analysis of the system, others, such as the needs of an unobtrusive measure, are necessary to ensure the comfort of the room occupants.

The number of sensors necessary to estimate the PMV value depends on the ambient complexity: the system must be scalable in order to be also applicable to complex environments, but a minimum of three sensors is required for the calculation of radiant temperature. Other sensors are embedded in clothes (i.e. one for every subject), therefore, the minimum solution depends on how many people we want to consider to set the climate.In order to fit the requirements in terms of accuracy and acquisition rate, we must consider the human thermal receptors' sensitivity and the environmental system dynamics:

The minimum temperature variation that can be detected by human bodies is approximately ± 0.3°C, more than 80% of people can detect variations within ± 0.6°C (Hsieh 2000)

The minimum RH variation detectable is near ± 3% RH (Meijer, and Van Herwaarden 1994).

Usually, temperature and humidity in a controlled environment take a large amount of time to change, at least some tens of seconds.

There is no reason to have much more accurate or fast sensors than the human body, and there are many commercial, low-cost solutions that satisfy these specifications in terms of accuracy, and the slow dynamics allow us to consider the data throughput from sensors to controllers negligible, avoiding critical requirements in terms of bandwidth and its related parameters, such as the battery power consumption. For fixed sensors we can have both a wired or a wireless connection, whilst a wireless connection is required for sensors positioned on the body of people, and, in order to extend battery life, there are well-known acquisition strategies to minimize power consumption.

3.3.3 Sensors

For a number of years, commercial temperature sensors have been made from semiconductors using a special bandgap junction. Working over a limited temperature range, they are simple, linear, accurate, and low-cost devices with many uses.

Many digital semiconductor temperature sensors will provide the required accuracy (within (± 0.5°C), but the main attraction of digital sensors is that they do not need further components (e.g. analog to digital converters) when connecting the sensor to the microprocessor (Table 3.1).

TABLE 3.1 Comparison of Four Different Digital Temperature Sensors

SENSOR	*SHT15*	*AD7414*	*DS1620*	*MAX6629*
Operating temperature	−40+123.8°C	−40+125°C	−55+125°C	−55+150°C
Accuracy	±0.3°C	±0.5°C	±0.5°C	±1°C
Resolution	0.01°C	0.25°C	0.5°C	0.0625°C
Supply current	30 A	900 A	1 mA	200 A
Operating voltage	2.4–5.5 V	2.7–5.5 V	2.7–5.5 V	3.0–5.5 V

TABLE 3.2 Comparison of Two Relative Humidity Sensors

SENSOR	*SHT15*	*HIH SERIES*
Operating humidity	0%–100% RH	0%–100% RH
Accuracy	±2% RH	±2% RH
Resolution	0.03% RH	NA (Analog output)
Supply current	30 A	200 μA at 5 V
Operating voltage	2.4–5.5 V	4.0–5.8 V

Most capacitive humidity sensors are composed of a thermoset polymer with three-layer capacitance and platinum electrodes. This type of sensor will produce capacitance relative to humidity, and the capacitance reading is converted to a voltage reading using on-chip circuitry. The voltage reading is linear to the relative humidity and is accurate to within 2% of the reading (Table 3.2.).

In this field of applications there are few devices that completely fit the specifications for an unobtrusive wearable measure, and some of them are also integrated with temperature sensors.

Figure 3.1 shows the performances of such wearable sensors in terms of variability. In fact, with the same subject and clothes but on different days, the measures show a good repeatability.

Analyzing data, it is also possible to observe the relation between modification of the RH and the activity of the people, and this consideration could open new scenarios of future studies to assess the possibility to indicate, and furthermore to estimate, in an automatic way, the changing of the metabolic activity (Figure 3.2).

Considering this late aspect, it will be possible to use the same technologies used to measure the RH and the temperature and, by means of improved models based on surface maps of these physical variables, to also deal with the complete automatic measure of the metabolic energy.

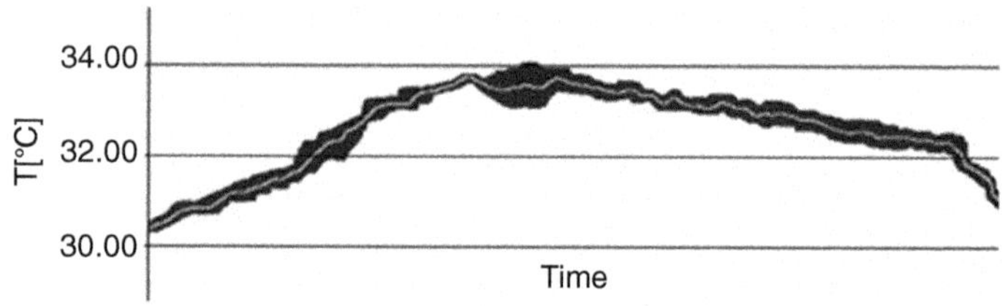

FIGURE 3.1 The figure represents the spreading of values obtained with three acquisitions on the same subject, in the same working conditions, and with the same clothes during the execution of the same activity. The white line shows the average of the three samples measured, while the black line represents the standard deviation.

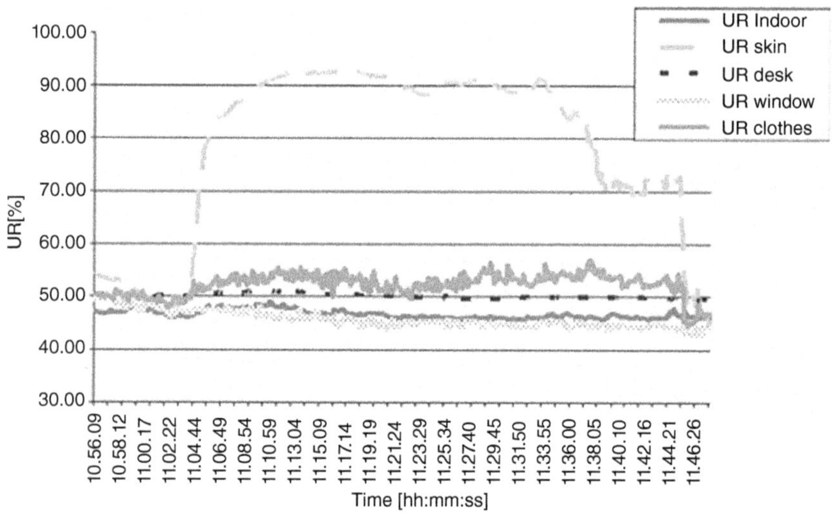

FIGURE 3.2 The figure represents the relative humidity measures behavior when the activity of the subject monitored increases (skin sensor).

3.4 CONCLUSION: THE NEW SCENARIOS OF HVAC

The paradigm presented in this chapter, and the measurements provided by the new technologies, offer solutions to solve Fanger's model. This model is suitable for applications in many locations where subjects spend a significant amount of time during their activities: the more complicated the environment, the greater is the number of transducers necessary to correctly estimate the PMV value. If we consider the concept and technologies introduced by domotics in recent years, it would be possible to imagine cost-effective, integrated, and diffused sensor networks. The main topic of how to apply the models and how they can be adapted to reach the complete autonomy of the control system has been discussed and shown in order to offer new perspectives. Many appropriate instruments are available, encouraging engineers, architects, and designers to move toward more dedicated personalized systems. These systems can follow the classical ones or, integrated with artificial intelligence, can be oriented to continuous learning and adaptation in order to satisfy some specific aims. Despite the potentiality of the described scenario, the overall performance will be considered in terms of accuracy of the estimation of the comfort index, sensitivity to the change (climate condition and subjects' activity), and quickness of response to these modifications. These parameters are important to project the integration with other systems and to satisfy the requirements using the specific actuators that buildings would integrate to control the climate. For example, they could address the reduction of consumption, costs, and pollution, whereas the physical parameters related to the psychophysical sensation estimation would be the same.

Despite the apparent expensive solutions, if we look at the worldwide research in ambient intelligence, it is clearly stated that this effort is worthwhile in the direction of a complete and personalized HVAC system based on the automatic estimation of thermal comfort. Accordingly, with the theoretical aspects indicated in the different models, this new approach could be used to develop new, more specific, and personalized devices. Furthermore, introducing embedded and smart solutions of disappearing computing, cognitive systems, nano and micro-technology will provide extremely cost-effective systems and would allow the integration of the environment with the people living and working inside it, in order to make homes, offices, and buildings more efficient, safe, and comfortable in general.

REFERENCES

Charles, K. E. (2003). Fanger's thermal comfort and draught models (IRC Research Report RR-162).

Di Rienzo, M., Andreoni, G., & Piccini, L. (2004). A wearable system for the unobtrusive measure of ECG. In *Proceedings of the International Federation for Medical and Biological Engineering*, Naples, Italy, July/August 2004.

Fanger, P. O. (1967). Calculation of thermal comfort: Introduction of a basic comfort equation. *ASHRAE Transactions*, *73*(2), III-4.1.

Hsieh, C. (2000). *Weather Net: Wireless office weather station*. University of Queensland.

International Standards Organization. (1997). ISO 7730: Moderate thermal environments—Determination of the PMV and PPD indices and specification of the conditions for thermal comfort.

Meijer, G. C. M., & Van Herwaarden, S. A. W. (1994). *Thermal sensors*. IOP.

Meyer, C. (2004). *Being comfortable with STAR-CD*. CD Adapco Dynamics, Spring, *21*.

FURTHER READING

Asim, N., Badiei, M., Mohammad, M., Razali, H., Rajabi, A., Chin Haw, L., & Jameelah Ghazali, M. (2022). Sustainability of heating, ventilation and air-conditioning (HVAC) systems in buildings: An overview. *International Journal of Environmental Research and Public Health*, *19*(2), 1016.

Solano, J. C., Caamaño-Martín, E., Olivieri, L., & Almeida-Galárraga, D. (2021). HVAC systems and thermal comfort in buildings climate control: An experimental case study. *Energy Reports*, *7*, 269–277.

Taheri, S., Hosseini, P., & Razban, A. (2022). Model predictive control of heating, ventilation, and air conditioning (HVAC) systems: A state-of-the-art review. *Journal of Building Engineering*, *60*, 105067.

Tejani, A., Yadav, J., Toshniwal, V., & Gajjar, H. (2022). Achieving net-zero energy buildings: The strategic role of HVAC systems in design and implementation. *ESP Journal of Engineering & Technology Advancements*, *2*(1), 39–55.

Wang, X., & Dong, B. (2024). Long-term experimental evaluation and comparison of advanced controls for HVAC systems. *Applied Energy*, *371*, 123706.

Working Clothing: Thermal Properties and Comfort Criteria 4

R. Ahasan and J. Salovaara

4.1 INTRODUCTION

Owing to unsuitable clothing, workers' physiological reactions are often reflected as deterioration of their work performances. A high level of protection by clothing can also severely impede heat exchange through the sweat evaporation. Thus, reliable measures to evaporative heat transfer through the design of suitable clothing should be developed. In this context, the thermal properties of clothing are reviewed, and corrective measures are emphasized that can match with workers' physiology, and on the effects of thermal balance, work performance and the comfort criteria.

Work clothing has a major contributing role to the workers' physiological response. It complicates the emission of body heat and causes physiological reactions through the increase of sweating, heart rate, and heat cramps, frequently impeding the worker's performance (Beshir 1994). Physiological changes and seasonal acclimation can lead to a thermal effect if clothing does

DOI: 10.1201/9781003512257-4

not properly match with the working environment. Working conditions also may be stressful due to incorrect or insufficient clothing, and for high metabolic rate of the workers. The effect of air temperature (T_a °C) and humidity is mainly reflected in skin temperature (T_{sk} °C) and in the clothing fabrics (Woodcock 1962). Higher humidity also causes excessive thermal stress more than in dry conditions (Boisvert *et al.* 1993; Candas *et al.* 1979). Sweat rate increases occur more with the humidity in clothing with higher insulation, and it can lead to an increase in T_{sk}. Thus, increasing concern has been growing among the concerned professionals about undesirable exposure to working clothing. Kenney *et al.* (1987) demonstrated workers' physiological limits in varying thermal environments. They illustrated the effects of moisture absorption in increased thermal load, integrating the coefficients of properties of heat exchange. Lotens (1993) derived a clothing model that handles most of the physical phenomena. Crockford (1994) illustrated special garments, which can reflect 40% of incoming radiation, and can reduce the convection loss by 25% if the clothing has 20% of maximum evaporative capacity (E_{max}). Sen and Dutta (1985) discussed heat radiation, and they proposed aluminum-foiled clothing for workers who are involved with furnace and steel millwork. Crockford *et al.* (1984) mentioned the design and suitability of protective clothing for workers. In a hot, humid climate, radiative heat load may be >400 W/m^2, and metabolic heat load may range >300 W/m^2, and thus, evaporative cooling is to be maintained (Kähkönen 1994). Physical work in hot, humid climate is heavy and characterized by static and dynamic loads of variable duration and intensity (Ahasan *et al.* 1996). Workers with low physical capacities are usually less able to tolerate heat, and they need good clothing. Environmental control measures are also important to reduce the risk of heat illness. In this regard, ISO 7933 (1989) and NIOSH (1986) have devised some principles that the work rate (or workload) should be compensated for by a reduction of heat load. Crockford (1994) added that lower E_{max} might limit the physical workload. By shortening the time of work exposure or changing the climate, thermal load can be reduced.

The limiting values (i.e. warning level of acclimatized persons) in accordance with ISO (International Standard Organization) and NIOSH (National Institute of Occupational Safety and Health, USA) standards provide the maximum level of safety for the majority of employees (Ilmarinen 1978). However, reduction of climatic and metabolic stress, and improvement of protective clothing remain a major challenge in the selection of fabrics. In this regard, several thermal indices and heat exchange model have been developed by Holmér (1995) to select clothing ensembles (Gagge *et al.* 1986). For this, individuals' metabolic condition, stature, sex, age and even social, religion and culture are very important considerations. Often clothing insulation is not adjusted to meet the requirements of the actual activity level. Thus, in the design and purchase of working clothing, local criteria should be considered.

Working clothing must be wearable that can balance evaporative cooling, and be readily available in the local market. The design has to be simple and easy to manufacture locally. It must be comfortable and likely to be functional for effective sweat evaporation through its fabrics. Many developing countries are poor in economy, where physical work combines with heat stress and causes acclimation through sweating. In this sense, thermal properties of tropical clothing (Ahasan *et al.* 1996) are very important that determine heat exchange rate, and as such, act as a resistance to heat loss through the evaporation of sweat. Therefore, the correct choice of working clothing depends on the heat stress index, permissible limit values, thermal properties and local climatic condition.

4.2 THERMAL PROPERTIES OF WORKING CLOTHING

Working clothing affects heat transfer from and to the human body mainly through the insulation and permeability of the fabrics. The subjects can feel pleasant due to optimum insulation and ventilation (i.e. concentration gradients) in the clothing across the textile layers (Nielsen *et al.* 1989). For instance, in tropical countries, heat radiation and humidity have a direct effect for which clothing has to be selected carefully. Solar heat radiation continues to be a serious heat problem for the workers in tropics, even moderate physical workload could be heavy and stressful in that situation (Sen 1965). Holmér (1985) also has reported the detrimental effects of moisture absorption in working clothing and insulation characteristics. Lotens and Havenith (1994) and Shapiro *et al.* (1995) discussed the effects of sweat absorption in working clothing.

Water absorption, transporting properties and drying rate of sweat are also evaluated (Nielsen 1992; Nielsen and Endrusick 1990; Nordon and David 1967) from the selected clothing ensembles. Evaporated sweat may condense and accumulate and thereby influence the distribution of humidity in working clothing (Madsen 1976). With evaporative heat transfer, it has been possible to reduce up to 50% using cotton clothing (Kenney *et al.* 1987). High relative humidity (RH) can also affect clothing layers, which is prevalent in the tropics. Ha and Tokura (1995) noted that sweats are significantly higher in polyester than cotton, and a wet fabric in direct contact with skin produces an unpleasant sensation as *chilly effect* or *clammy feelings*. Radiative energy in tropical climate is an additional heat load while wearing color-absorbing and reflecting clothing (Lotens and Pieters 1995). That is why the thermal properties of clothing ensembles and fabric materials are to be given priority.

4.2.1 Clothing Insulation

Insulation characteristics are important for working clothing to select proper clothing ensembles. It plays a vital role in moisture distribution and sweat absorption and has a leading influence on the comfort of clothing (Fanger 1972; Givoni and Goldman 1972; Holmér and Elnäs 1981; ISO-9920 1993; Nielsen and Toftum 1992). Fabric surface and its thickness, i.e. amount of air trapped inside fabrics, mainly determine thermal insulation of clothing (or Clo). Clo (e.g. thermal insulation or resistance of fabrics are necessary to maintain the thermal balance between T_{sk}, textile materials and T_a for different clothing ensembles can be found from DIN standard 33403 (1980), Olesen and Nielsen (1983) and the ISO's draft standards. It can control the heat exchange rate between the skin and the environment. Clothing insulation is varied to the individual's physiological response, and to the combination of work-load, RH, T_a air velocity (V_a) and clothing ensembles. Olesen (1982) have evaluated Clo by movable thermal manikin on the effects of body posture and activity level.

Permeability of clothing ensembles indicates the ability of water vapor to pass through the fabrics (i.e. ventilation). High permeability means more sweat will be allowed to evaporate, and high ventilation can enhance cooling of the skin through sweat evaporation. Light and moderate insulated, permeable with good ventilated clothing (long sleeve shirt, trouser, socks and shoes having 0.6 Clo) are recommended where no risk of high contamination (Holmér and Elnäs 1981). Fabric thickness and clothing layers also reflect clothing insulation (Larose 1947). Typical clothing ensembles (Holmér and Elnäs 1981; ISO-9920 1993, McCullough *et al.* 1989; Oohori *et al.* 1985) are listed in Table 4.1, which are comparatively available.

TABLE 4.1 Clothing Insulation and Permeability Index of Clothing Ensembles

CLOTHING ENSEMBLES	*CLO-VALUES*		*PERMEABILITY INDEX*		*TER*
($R_{T,EVP}$)	L_{CL}	L_T	I_M	I_{EL}	*(PA M²/W)*
Cotton shirt, trouser, underwear, shoes	0.65	1.25	0.39	0.32	30
T-shirt, underwear, socks, shoes	0.84	1.41	0.39	—	34
Shirt, pant, underwear, belt, socks	1.00	1.51	0.30	0.27	44
Aluminized coat, CST uniform	1.36	1.89	0.33	—	53
Aluminized coat (midnight) OS-uniform	1.74	2.19	0.30	0.27	68
Chemical protective clothing (Goretex)	1.40	1.90	—	—	180

Note: I_T is total insulation; i_m and i_{cl} are permeability index and/or boundary layer, and clothing layer, respectively under the standard conditions. $R_{t,evp}$ is the total evaporative resistance.

4.2.2 Clothing Ensembles and Fabric Materials

Clothing ensembles and fabric materials have a significant impact on thermal properties and comfort. Nielsen (1993) emphasized on dissipation of metabolic heat that can be reduced up to 25% depending on proper fabrics. High humidity and solar radiation cause excessive sweating and, thus, Hall (1971) developed a simplified method for clothing ensembles. Lotens and Pieters (1995) evaluated the radiation effect on two-layer clothing (Figure 4.1).

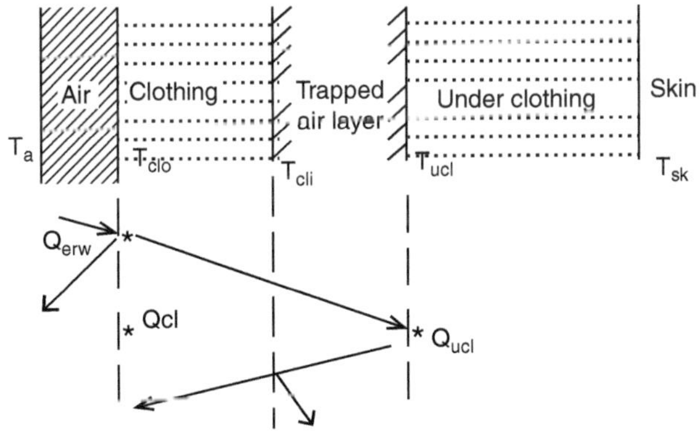

FIGURE 4.1 Radiation effect on clothing ensemble consisting of outer garment and under clothing.

Heat is absorbed first in the outer layer Q_{cl} and partly transmitted to underclothing Q_{ucl}, until the radiation energy is completely distributed over the clothing layers and then to the environment Q_{env}. The resultant temperature in the fabrics (T_{ucl} for underclothing, T_{cli} for inside of outer clothing and T_{clo} outside of outer clothing) depends on the distribution of Q_{env} over the clothing layers, and on the heat resistance of fabric layers and the boundary conditions (T_{sk}, T_a). Absorption properties of working clothing show a greater or equal transit in T_{sk} and may have beneficial effect on thermal sensation. During absorption, vapor concentration moves upward (Figure 4.2) along the line, and in a downward direction for desorption (Lotens and Havenith 1994).

Sweating is related with fabrics and clothing ensemble are also needed to be formulated from the required sweat rate (SW_{req}) and Clo which depend on physical properties of fabrics (Table 4.2): thickness (density, course, number inch),

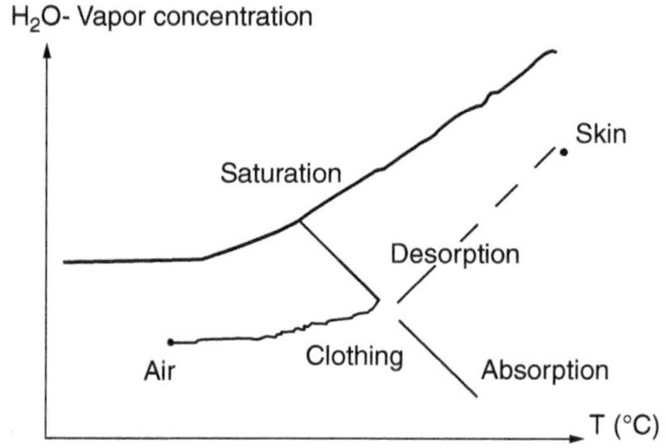

FIGURE 4.2 Effects of H_2O vapor absorption in clothing, at skin and in surrounding air.

TABLE 4.2 Physical Characteristics of Textile Materials [ISO-5084 (1977); ISO-5085 (1987); CAN2-4.2M77 (1977)]

	THICKNESS	*SP.WT*	*TR*	*AP*	*WVR*	*HFTR*
FABRICS	*(MM)*	*(G/M²)*	*(KM²/W)*	*(L/M²/S)*	*(MM AIR)*	*(%)*
Cotton	1.52	153	0.028	1741	*	62
Polyester	0.73	280	0.016	104	1.4	69
Polyester/cotton mixed	2.40	210	0.058	983	2.3	55

specific weight (sp. wt.), thermal resistance, water absorbency; thermal conductivity; moisture transfer, air permeability (AP), water vapor resistance (WVR) and heat flow transmission rate (HFTR).

Absorption capacity for the actual wearer has seldom been quantified, but Nordon and David (1967) developed a numerical method including the rate of equilibrium between fabrics and surrounding air. To simulate clothing ensembles, Havenith and Lotens (1994) developed a computer model based on the geometry of clothing and the activity of the worker. They focused on the prediction of thermal effect on heat balance through the process of vapor absorption in different fabric layer. Richardson (1965) reported the effects of saturated sweat absorption in clothing ensembles, using a thermal manikin. More recently, De Dear *et al.* (1989) reported the thermal effects, moving from one environmental humidity to another with absorbing and non-absorbing clothing.

4.2.3 Heavy Protective Clothing for Working in Dangerous Conditions

Heavy protective working clothing is important for pesticide sprayers, in chemical plants (or waste disposal), food industries and in nuclear power plants (Mihal 1981). Flameproof clothing is used for fire brigades, and nuclear, biological and chemical (NBC) clothing is used in particular situations to reduce the danger of work hazards. Heavy protective over-garments and water-repellent fabrics are also used in heavy protection; however, this clothing is very expensive. Protective clothing with microporous and ventilated fabrics can significantly improve heat exchange and reduce physiological strain. An adjustable valve must be efficient in NBC clothing to wash out most of the metabolic heat. NBC clothing poses heavy constraints to work, even with moderate work, particularly with high humidity. Therefore, a proper schedule of work and better ventilation should be provided (Ohnaka 1993). Protective clothing should design that heat tolerance in terms of dynamic changes to inside body temperature must not be above the tolerance level, as well as to improve work tolerance and can protect against chemical liquids, dusts, aerosols and poisonous substances. For better comfort and performance, the temperature of the space between skin and clothing layer (T_L) should be monitored carefully. Cortili *et al.* (1996) found a reliable match of T_L between 36.2 and 36.4°C (when V_a = 0.1 and 0.5 respectively) that may valid mainly in the tropical climates (T_a = 35°C, RH = 85%).

4.2.4 Typical Working Clothing

Typical clothing and its model depend on multiple factors such as sex, age, health, stature and body composition. Even social status, culture, religion and region can affect thermal responses. For instance, Thai female workers are shorter than Tanzanian women, and their clothing is not similar. Females in SAARC (Southeast Asian Association for Regional Co-operation: India, Nepal, Bangladesh, Pakistan, Bhutan; Maldives, Sri Lanka, etc.) countries use *sari* (one piece of long cloth that is to be trapped and cover all the whole body, blouse and petticoats). *Sallower* and *kamiz* (long-legged trousers and shirts) are popular for young women. *Lungis* (hollow type long legged clothing) and sandals are very common in men's wearing in these regions. In the big cities, working clothing are usually pants, shirts, aprons and shoes. Women in the ASEANS (Association of Southeast Asian Nations) use pants, skirts and tops. African and Arab women use hollow-type dresses. Owing to religion, shorts and light clothing are forbidden in some Muslim countries (e.g. Malaysia, Saudi Arabia), and even there are State rules that clothing must be worn for complete cover of all sexual parts. However, working clothing should be selected so that it can reduce physiological and climatic stress. Generally, cotton, nylon and polyester are used as light working clothing. Traditional or regular working clothing is generally light dresses that include trousers, T-shirt, sandals or canvas shoes (0.5–0.7 Clo).

ACGIH (1991) devised summer clothing (0.6 Clo); Ramsey (1978) suggested normal clothing (0.5–1.0 Clo); Goldman (1981) added regular clothing (0.5–0.7 Clo); and Bernard *et al.* (1986) preferred cotton coveralls (0.7–1.0 Clo). Jogging shoes, socks, shorts, single jacket or T-shirt are athletic clothing. Aluminized garments are used for workers in the steel and re-rolling mills to protect them from radiant heat. Further, proper working clothing should be considered from the available fabrics, which are cheaper and thermally matched with metabolic reactions, because ISO and NIOSH recommended clothing is not easily suitable and cheap in some circumstances.

4.3 COMFORT CRITERIA

Comfort criteria are very important for the design of working clothing. To assess proper working clothing, thermal properties of fabrics, and the effects of metabolic reactions should be considered. Feeling of clothing discomfort also depends on the type of tasks, jobs, work environment and body heat balance on clothing ensemble (Hafez and Beshir 1987; Wyon *et al.* 1975).

ISO standards have evaluated thermal sensation and discomfort levels using various clothing ensembles. Galbraith *et al.* (1962) evaluated comfort sensation while wearing cotton clothing, for example light gray, pure, white or mixed cotton, linen or polyester (65% cotton plus 35% polyester or linen) are also preferred. Ramsey (1978) described physiological responses for general guidelines, but this approach does not contribute to the heat balance equation. Fanger (1972) developed mean thermal sensation on clothing effects. Neilsen *et al.* (1989) predicted comfort level using subjective and physiological rating scales. Havenith and Lotens (1984) compared semipermeable and impermeable garments to heat dissipation and discomfort of thermal sensation. Hensel (1981) showed the relationship between thermal discomfort and skin wettedness by various ensembles. For thermal comfort, seasonal variation in T_a, V_a heat radiation and humidity are to be selected properly (Vokac *et al.* 1976) because physiological change and seasonal acclimation can lead to thermal effect if activity and clothing ensembles do not match properly.

Working clothing effect on climatic and physiological parameters is to be integrated for acceptable thermal sensation (Sprague and Munson 1994). In general, subjects in tropical climates feel comfortable using light clothing due to solar heat and higher humidity (e.g. RH $\cong$ 70%–85%). Moisture absorption is generally regarded as a prime factor in thermal comfort that depends on fabric rigidity, wicking, clinging, liquid buffering, drying ability and other thermal properties of working clothing. Condensation of evaporated sweat is highly relevant with comfort criteria (Gonzalez *et al.* 1986); and the color of clothing and the wavelength of radiation effect on clothing are very important (Nielsen 1990).

4.4 DISCUSSION

Recommendations for proper working clothing should rely on the understanding of the physiological effects of heat exposure to clothing by environmental control that can enhance safety and work performance. Kenney *et al.* (1986) emphasized acclimatization control advocated by specific measures (using natural and cross ventilation, heat barriers, reduction of physical work, rest breaks or provision of drinking water). Local contexts are to be applied for physiological adaptations with working clothing and climatic variables, because threshold limits (TLV) or recommended standards as national legislation in tropics are yet to be established. In most cases, workers or entrepreneurs know those parameters, but they do have some sort of thermal sensation from which clothing can be selected. Acceptable working clothing must be chosen by physiological and climatic conditions that match properly. The workers are to be advised to wear proper clothing and are trained through the metabolic heat reduction scheme (Ramsey 1995) and overall environmental control) by certain control (NIOSH 1986; ASHRAE 1992; ISO-7730 1993; Olesen and Madsen 1995). To protect from the ill-effects, TLV or certain standards are important for clothing control that keep the body heat balance and that can bring a massive advantage to workers' health, safety, efficiency and productivity. Following standards are important for recommendations: general guidelines for metabolic heat exposure limits, and selection of normal clothing, unacclimatization and environmental control (Kodak 1983; NIOSH 1986); appropriate Clo and design of clothing (Ramsey 1978, 1995; Bernard *et al.* 1991); modification of wet-bulb glove temperature (WBGT) index and correction factors for various clothing (ISO-7933 1989; ACGIH 1991); estimation of thermal insulation of clothing (ISO-9920 1993); evaluation of moderate thermal work environment, and prediction of thermal sensation and discomfort (ISO-7730 1993); guidelines for evaluation of thermal strain (ISO-9886 1992); subjective judgement of thermal sensation (ISO DIS-10551 1993) and ergonomics of thermal environment (ISO CD-12894 1993). This established standard means a situation or measures above which work should not be resumed wearing working clothing. According to ASHRAE (1977), comfort range is ~23–27°C. Fanger (1970) added that clothing having 0.5–0.7 Clo is better for sedentary work during lower V_a (>0.3 m/s). Sen (1982) advised that T_{room} at 23.5°C is comfortable at lower humidity levels (e.g. 50% RH). He found a comfort zone in the Indian climate to be at 26°C in corrective effective temperature for those engaged in light or moderate work. Air-conditioning systems are used in many commercial offices, market complexes or banks where a thermal comfort zone is taken into consideration.

Common ergonomics approaches are the arrangement of proper clothing materials with less thermal impact (Epstein *et al.* 1986). Fabric thickness affects insulation and can influence mean skin temperature because of vapor diffusion and moisture transportation from skin to the clothing ensemble (Bakkevig and Nielsen 1994). For example, a fishnet textile allows more air to sweep the sweating and at the same time can act as effective ventilation (Nielsen and Endrusick 1990). Normally cotton, gabardine and mixed fabrics (cotton, gabardine, linen) are a benefit in the tropics (Kogi and Sen 1987). They added that fabrics should be flame-resistant, flexible and washable. Foot and underwear are to be suitable. Sandals are extensively used for light wearing, especially in tropical developing countries, but to ensure more safety, canvas shoes are recommended. Soft rubber or plastic shoes can be used for moderate work. Gym shoes and boots are recommended for heavy work where protection (e.g. from burning or cutting) of feet is vital. Sweat accumulation is generally higher in underwear that can be affected by fiber types. Excessive sweat may feel uncomfortable, and thus underwear clothing (i.e. fabric construction) should be selected carefully (Bakkevig and Nielsen 1994).

It has been evident that most of the protective clothing severely restricts evaporative heat loss and increases thermal strain, but microporous fabric can significantly protect against chemical liquids, dusts or aerosols, and impermeable fabrics can protect against gases and vapors. NBC clothing restricts heat loss by evaporation when radiation is direct or omnidirect. So, adjustable valves for air outlet must be provided to transmit sweating and heat through the clothing fabrics. Smith *et al.* (1994) emphasized on TLV and exposure time that physical fatigue can be normalized. Madsen (1976) suggested that workers in cold storage, slaughtering house or in a fish processing factory need multi-layered clothing ($I_T = 1.20$ Clo). He added that the outer layer should be cotton, the middle layer a mixed polyester and cotton, and the inner layer could be pure cotton. Makambaya (1994) added special working clothing (outer windproof and waterproof garments) for high wind and rain. Hygroscope and damped humidity-gradient fabrics are also suitable for transferring saturated sweat from the skin. Anyway, open and thin garment knits are preferred, especially for tropical countries.

4.5 CONCLUSION

Most of the entrepreneurs cannot afford costly clothing for their workers; however, working clothing should be selected on the basis of thermal properties (e.g. suitable and adjustable Clo), available in the market and long-lasting with reasonable price. Generally, cotton fabrics are the cheapest and comfortable in the tropics. NBC and/or heavy protective clothing are used in the cold climate, which is a bit expensive. But modifications of heat exchange should be adjusted carefully whatever the working condition is. Moreover, it has been evident that impaired heat exchange with heavy protective clothing contributed to most of the physiological strain. The workers' general opinion is that protective clothing is sometimes more of a handicap than a protection. Therefore, it is important to know a reliable indicator of thermal load in a wide range of different combinations of physiological and climatic variables on this clothing.

Physiological effects and the significance of clothing on tropical humans are not yet completely known. In this regard, subjective assessment is to be evaluated where local influences, such as asymmetric radiation, draught, vertical rays or direct radiation exist. Moreover, control of heat exposure must satisfy national health and safety standards (or legislation); however, many countries are yet to establish legislation. Clothing must be worn to reduce the thermal impact as well as to protect against physical and environmental work hazards. Thus, thermal properties of clothing should be evaluated on the effects of local population (i.e. on physiological adaptation with local climate) that can match with thermal comfort sensation. The practitioners normally estimate metabolic rates following heat stress standards, but thermal comfort reflect on the integrated state of thermoregularity system, where T_{sk} and internal body temperature (T_{body}) must be integrated together. Metabolic rate and clothing values must be keyed in by ergonomist to design a proper choice of working clothing. So, careful control of exposure duration, thermal indices (T_a, RH, V_a, radiant heal flux, solar radiation, etc.) and metabolic factors (sex, age, obesity, health status, physical fitness, skin temperature and wetness, etc.) are important to be considered (Lee 1980). Wearing NBC clothing results in increased respiratory strain and elevated T_{body}, and inspiratory or expiratory effect. But distributing it in multilayered ensembles can reduce sweat accumulation. Sweat may not evaporate completely in these clothing, but if inner layers are absorbing in nature then it will perhaps be benefited. Outer layers must be easy to open to allow sweat be exposed to the air. Latent heat of condensed sweat must be available for conduction to clothing surface and to the environment. Under certain conditions, evaporative sweat may condense before escaping to the environment, giving off its heat to the surrounding clothing layers. The concept of basic insulation (I_{cl}) is valid for a limited range of ensembles and thermal conditions. In this regard, the effect of condensation evaporation (efficiency and conductive heat exchange) requires further experimental work along with consideration of work posture, body motion and wind effects.

Recent modifications by ISO standards are rational but not promising for low-cost ensembles. Sweat rate prediction by the WBGT index may not reflect the actual heat stress for clothing design if there is high T_r, T_a or V_a.

It also seems that the climate conditions in thermal work environment for which ISO standards are applicable are too narrow in the tropical climate. The estimating errors of I_{cl} may not be always correct in evaluating the thermal environment and the selection of proper clothing. Tropical countries are experienced with heavy physical work, and wearing protective clothing can increase sweating and vapor concentration which results in uncomfortable feelings. In most of the developing countries workers are less educated and lack knowledge about work hazards. Environmental control should relate to the use of proper clothing to enhance safety and work performance. A model should be derived for standard clothing, especially for SAARAC, ASEANS, African, Arab and other tropical countries that handle most of the physical phenomena influencing heat transfer criteria. Hence, further R&D should focus on the ability to interact sweat transportation, heat dissipation, and on how such effects can modify Clo for cheaper and available clothing. Standard clothing recommended by ISO, NIOSH or ASHRAE is also very expensive and, therefore, ergonomics of low-cost clothing should be a great achievement.

REFERENCES

ACGIH (1991). *Threshold limits values for physical agents in the work environment and heat stress.* Cincinnati: American Conference on Governmental and Industrial Hygienists.

ASHRAE (1977). *Physiological principles, comfort, and health.* American Society of Heating, Refrigeration, and Air Conditioning Engineers.

ASHRAE (1992). *Thermal environmental conditions for human occupancy.* American Society of Heating, Refrigeration, and Air Conditioning Engineers.

Bakkevig, M. K., & Nielsen, R. (1994). Impact of wet underwear on thermoregulatory responses and thermal comfort. *Ergonomics, 37*(8), 1375–1389.

Bernard, T. E., Kenney, W. L., Balint, L., & O'Brien, J. F. (1986). *Heat stress management program for nuclear power plants.* Palo Alto: Electric Power Research Institute.

Bernard, T. E., Kenney, W. L., Hanes, L. F., & O'Brien, J. F. (1991). *Heat stress management program for power plants.* Palo Alto: Electric Power Research Institute.

Beshir, M. Y. (1994). Thermal work environment in Africa. *African Newsletter on Occupational Health and Safety, 4*(1), 3.

Boisvert, P., Nakamura, K., Shimai, S., Candas, V., & Tanaka, M. (1993). Sweat rate and body temperature during exercise in dry and humid environments. In R. Nielsen & K. Jorgensen (Eds.), *Advances in industrial ergonomics and safety V* (pp. 399–409). London: Taylor & Francis Group.

CAN2-4.2-M77 (Method-49). (1977). Method for the test of resistance of materials to water vapor diffusion (Control-dish method).

Candas, V., Libert, J. P., & Vogt, J. J. (1979). Influence of air velocity and heat acclimation on human skin wettedness and sweating efficiency. *Journal of Applied Physiology, 47*(6), 1194–1200.

Cortili, G., Mognoni, P., & Saibene, F. (1996). Work tolerance and physiological response to the thermal environment wearing protective clothing. *Ergonomics, 39*(4), 620–633.

Crockford, G. W. (1994). Work wear and protective clothing for hot workplaces in tropical countries. *African Newsletter on Occupational Health and Safety, 4*(2), 16–17.

Crockford, G. W., Sen, R. N., & Spencer, J. (1984). The design of work wear and protective clothing for use in the tropics. *Indian Journal of Physiology and Allied Sciences, 38*(3), 125–133.

De Dear, R. J., Knudsen, H. N., & Fanger, P. O. (1989). Impact of air humidity on thermal comfort during step changes. *ASHRAE Transactions, 95*(Pt 2), 336–350.

Epstein, Y., Shapiro, Y., & Brill, S. (1986). Comparison between different auxiliary cooling devices in a severe hot and humid climate. *Ergonomics, 29*(1), 41–48.

Fanger, P. O. (1970). *Thermal comfort: Analysis and applications in environmental engineering.* New York: McGraw-Hill.

Fanger, P. O. (1972). *Thermal comfort.* New York: McGraw-Hill.

Gagge, A. P., Fobelets, A. P., & Berglund, L. G. (1986). A standard predictive index of human response to the thermal environment. *ASHRAE Transactions, 92*, 709–731.

Galbraith, R. L., Werden, J. E., Fahnestock, M. K., & Price, B. (1962). Comfort of subjects clothed in water-repellent clothing and Orion suits. *Textile Research Journal, 32*(3), 236–242.

Givoni, B., & Goldman, R. E. (1972). Predicting rectal temperature response to work, environment, and clothing. *Journal of Applied Physiology, 32*(6), 812–822.

Goldman, R. F. (1981). Prediction of heat strain: Revised 1979–1980. In F. N. Dukes-Dobos & A. H. Hensel (Eds.), *Proceeding of a NIOSH workshop on recommended heat stress standards (No. 81–108).* Cincinnati: National Institute for Occupational Safety and Health.

Gonzalez, R. R., Breckenridge, J. R., Levell, C. A., Kolka, M., & Pandolf, K. B. (1986). Efficacy of heat exchange using wettable cover over chemical protective garments. In ASTM (Ed.), *ASTM standards* (pp. 515–534).

Ha, M., & Tokura, H. (1995). Effect of two kinds of clothing made from hydrophobic and hydrophilic fabrics and local sweating rates at 37°C. *Ergonomics, 38*(11), 1445–1455.

Hafez, H. A., & Beshir, M. Y. (1987). Effects of thermostat temperature on human performance. In S. S. Asfour (Ed.), *Trends in human factors IV* (pp. 325–332). Amsterdam: Elsevier Science.

Hall, S. A. (1971). Heat stress in outdoor manual workers in East Africa. *Ergonomics, 14*(1), 91–94.

Havenith, G., & Lotens, W. A. (1984). *What actually is the advantage of semi-permeable over impermeable rainwear? Report IZF-6.* Soesterberg: TNO Institute for Perception.

Hensel, H. (1981). *Thermoreception and temperature regulation.* London: Academic Press.

Holmér, I. (1985). Heat exchange and thermal insulation compared in woollen and nylon garments during wear trials. *Textile Research Journal, 55*(8), 511–518.

Holmér, I. (1995). Protective clothing and heat stress. *Ergonomics, 38*(1), 166–182.

Holmér, I., & Elnäs, S. (1981). Physiological evaluation of resistance to evaporative heat transfers. *Ergonomics, 24*(1), 63–74.

Ilmarinen, R. (1978). *Influences of clothing ensembles on several physiological parameters during work in hot and humid environments [Dissertation].* Deutsche Sporthochschule, Cologne.

ISO CD-12894. (1993). *Ergonomics of thermal environment — Medical supervision of individuals exposed to hot and cold environments.* Geneva: International Standard Organization.

ISO-5084. (1977). *Textiles — Determination of thickness of woven and knitted fabrics.* Geneva: International Standardization Organization.

ISO-5085. (1987). *Textiles — Heat retention properties fabrics by means of heating body.* Geneva: International Standard Organization.

ISO-7730. (1993). *Moderate thermal environment — Determination of PMV and PPD indices and specifications of the conditions of thermal comfort.* Geneva: International Standard Organization.

ISO-7933. (1989). *Hot environments — Analytical determination and interpretation of thermal stress using calculation of required sweat rate.* Geneva: International Standard Organization.

ISO-9886. (1992). *Evaluation of thermal strain by physiological measurements.* Geneva: International Standard Organization.

ISO-9920. (1993). *Ergonomics — Estimation of thermal characteristics of clothing ensembles.* Geneva: International Standard Organization.

Kähkönen, E. (1994). Thermal work environment and ISO standards. *African Newsletter on Occupational Health and Safety*, *4*(1), 4–6.

Kenney, W. L., Lewis, D. A., Anderson, R. K., & Kamon, E. (1986). A simple test for prediction of relative heat tolerance. *American Industrial Hygiene Association Journal*, *47*(4), 203–206.

Kenney, W. E., Lewis, D. A., Hyde, D. E., Dyksterhouse, T. S., Armstrong, C. G., Fowler, S. R., & Williams, D. A. (1987). Physiologically derived critical evaporative coefficients for protective clothing ensembles. *Journal of Applied Physiology*, *63*(3), 1095–1099.

Kodak. (1983). *Ergonomic design for people at work* (Vol. *1*). New York: Eastman-Kodak.

Kogi, K., & Sen, R. N. (1987). Third World ergonomics. In D. J. Oborne (Ed.), *International reviews of ergonomics* (Vol. *1*, pp. 77–118). London: Taylor & Francis Group.

Larose, P. (1947). The effects of wind on thermal resistance of clothing with special reference to protection by coverall fabric of various permeability. *Canadian Research Journal*, *25*, 169–190.

Lee, D. H. K. (1980). Seventy-five years search for heat index. *Environment Research*, *22*(2), 331–356.

Lotens, W. A. (1993). *Heat transfer from humans wearing clothing*. Soesterberg: TNO Institute for Perception.

Lotens, W. A., & Havenith, G. (1994). Effects of moisture absorption in clothing on the human heat balance. *Ergonomics*, *38*(6), 1092–1113.

Lotens, W. A., & Pieters, A. M. J. (1995). Transfer of radiation heat through clothing ensembles. *Ergonomics*, *38*(7), 1132–1135.

Madsen, T. L. (1976). Thermal manikin for measuring the thermal insulation capacity for human clothing (Paper No. 48). Thermal Insulation Laboratory, Technical University of Denmark.

Makambaya, S. (1994). Work in cold environment. *African Newsletter on Occupational Health and Safety*, *4*(1), 4–6.

McCullough, E. A., Jones, B. W., Zbikowski, P. J., Arpin, E. J., Konz, S. A., & Rohles, F. H. (1989). *ASHRAE Transactions*, *89*(1), 327–352.

Mihal, C. P. (1981). Effects of heat stress on physiological factors for industrial workers performing routine work and wearing impermeable vapor-barrier clothing. *American Industrial Hygiene Association Journal*, *42*(2), 97–103.

Nielsen, B. (1990). Solar heat loads: Heat balance during exercise in clothed subjects. *European Journal of Applied Physiology*, *60*(5), 452–456.

Nielsen, B. (1992). Heat stress causes fatigue: Exercise performance during acute and repeated exposure to hot and dry environment. *Medicine and Sport Science*, *34*, 207–217.

Nielsen, B. (1993). Evaporation of sweat from dressed man. In R. Nielsen & K. Jorgensen (Eds.), *Advances in industrial ergonomics and safety V* (pp. 429–434). London: Taylor & Francis.

Nielsen, R., & Endrusick, T. L. (1990). Thermoregulatory response to intermittent exercise is influenced by knit structure of underwear. *European Journal of Applied Physiology*, *60*(1), 15–25.

Nielsen, R., Gavhed, D. C. E., & Nilsson, H. (1989). Thermal function of a clothing ensemble during work: Dependency on inner layer fit. *Ergonomics*, *32*(12), 1581–1594.

Nielsen, R., & Toftum, J. (1992). *Udvikling af beklædning til kolde arbejdsmiljøer.* Copenhagen: Arbejdsmiljøfonden.

NIOSH. (1986). *Criteria for a recommended standard: Occupational exposure to hot environments (DDHS 86-113).* Washington, DC: National Institute for Safety and Health.

Nordon, P., & David, H. G. (1967). Coupled diffusion of moisture and heat in hygroscopic textile materials. *International Journal of Heat and Mass Transfer*, *10*(5), 853–866.

Ohnaka, M. (1993). Critical size of the nucleation zone of earthquake rupture inferred from immediate foreshock activity. *Journal of Physics of the Earth*, *41*(1), 45–56.

Olesen, B. W. (1982). *Thermal comfort* (Vol. 2). Bruel & Kjaer.

Olesen, B. W., & Madsen, T. L. (1995). Measurements of the physical parameters of the thermal environment. *Ergonomics*, *38*(1), 138–153.

Oohori, M., Ueda, S., & Watanabe, Y. (1985). Changes in Surface Condition of High Chrome Work Rolls in Hot Rolling Mill. *Transactions of the Iron and Steel Institute of Japan*, *25*(6).

Rabiul Ahasan, M., Väyrynen, S., & Kirvesoja, H. (1996). Physical workload analysis among small industry activities using postural data. *International Journal of Occupational Safety and Ergonomics*, *2*(1), 27–34.

Ramsey, J. D. (1978). Abbreviated guidelines for heat stress exposure. *American Industrial Hygiene Association Journal*, *39*(6), 491–495.

Ramsey, J. D. (1995). Task performance in heat: A review. *Ergonomics*, *38*(1), 154–165.

Richardson, D. L. (1965). *Study and Development of Materials and Techniques for Passive Thermal Control of Flexible Extravehicular Space Garments* (Vol. *65*). Aerospace Medical Research Laboratories.

Sen, H. K. (1965). Evaporation of the Solar Corona into Interplanetary Space (No. 103). Air Force Cambridge Research Laboratories, Office of Aerospace Research, United States Air Force.

Sen, R. N. (1982). Certain ergonomic principles in the design of factories in a hot climate. In *Proceedings of the International Symposium on Occupational Safety, Health and Working Conditions in Relation to Transfer of Technology to the Developing Countries* (pp. 123–147). Geneva: International Labour Organization (ILO).

Sen, R. N., & Dutta, N. C. (1985). An ergonomic design of low-cost protective work clothing for the furnace workers. In *Proceedings of the International Symposium on Ergonomics in Developing Countries*, Jakarta, 18–21 November. Geneva: International Labour Organization (ILO).

Shapiro, Y., Moran, D., & Epstein, Y. (1995). Validation and adjustment of mathematical prediction model for sweat rate response to outdoor environmental conditions. *Ergonomics*, *38*(5), 981–986.

Smith, G., Bishop, P., Beaird, J., Ray, P., & Smith, J. (1994). Physiological factors limiting work tolerance in chemical protective clothing. *International Journal of Industrial Ergonomics*, *13*(2), 147–155.

Sprague, D. H., & Munson, D. M. (1994). A complete ensemble method for estimating thermal insulation values of clothing. *ASHRAE Transactions*, *80*(1), 120–129.

Vokac, Z., Köpke, V., & Keul, P. (1976). Physiological responses: Thermal humidity and comfort sensations in wear with cotton and polypropylene vests. *Textile Research Journal*, *46*(1), 30–38.

Woodcock, A. H. (1962). Moisture transfer in textile systems. *Textile Research Journal*, *32*(7), 628–633.

Wyon, D. P., Fanger, P. O., Olesen, B. W., & Pedersen, C. J. (1975). The mental performance of subjects clothed for comfort at two different air temperatures. *Ergonomics*, *18*(4), 359–374.

FURTHER READING

Mocan, A., Gaureanu, A., Szabó, G., Mrugalska, B. (2022). Arguments for emerging technologies applications to improve manufacturing warehouse ergonomics. In: Draghici, A., Ivascu, L. (Eds) *Sustainability and innovation in manufacturing enterprises: Advances in sustainability science and technology*. Springer, Singapore. https://doi.org/10.1007/978-981-16-7365-8_5

Kim, D. H., & Lee, J. Y. (2023). Heat strain while wearing pesticide protective clothing in hot environments: effects of textile physical properties and ambient humidity. *International Journal of Industrial Ergonomics*, *93*, 103388.

Teyeme, Y., Malengier, B., Tesfaye, T., Ciesielska-Wrobel, I., Haji Musa, A. B., & Van Langenhove, L. (2021). A review of contemporary techniques for measuring ergonomic wear comfort of protective and sport clothing. *Autex Research Journal*, *21*(1), 32–44.

Warthmann, A., Kohri, I., Ozeki, Y., Nagano, H., & van Treeck, C. (2024). Equivalent contact temperature (ECT) for personal comfort assessment–analytical description and definition of comfort limits. *Ergonomics*, *67*(2), 207–224.

Watson, C., Troynikov, O., Kuklane, K., & Nawaz, N. (2021). Industrial workwear for hot workplace environments: thermal management attributes. *International journal of biometeorology*, *65*(10), 1751–1765.

5 Work in Extreme Environments: Physiological Fundamentals

M. Kolich

5.1 INTRODUCTION

Through work, humans are subjected to a wide variation in environmental conditions. Not only are they asked to adjust to this variation, but they are also asked to perform at a high level. The central problem for humans, therefore, is to maintain an internal steady state despite an often harsh and fluctuating environment.

The purpose of this chapter is to present a summary concerning the fundamentals of human thermoregulation. It is, by no means, intended to provide a complete picture of any one topic in particular. It is meant as background for the follow-up papers dealing with the performance aspects of work in extreme environments and future research directions.

DOI: 10.1201/9781003512257-5

5.2 FUNDAMENTALS OF HUMAN THERMOREGULATION

Despite millions of years of evolution, humans have adapted to varying climatic conditions principally by altering the conditions under which they live rather than through physiological adaptation. They build shelters, wear appropriate clothing, and provide heating and air conditioning. Without these things, humans could only survive in a narrow zone along the equator where the ambient temperature is between 28°C and 30°C.

Thermoregulation is achieved by balancing the two main factors that determine body temperature—the metabolic heat produced and the rate of heat loss. In humans, the goal is to maintain the temperature of the deep body tissues (the core temperature) within a narrow range. Core temperatures above 39.5°C are disabling, and over 42°C, they are usually fatal. The lower acceptable limit is 35.5°C, and 33°C marks the onset of cardiac disturbances. Further drops in core temperature are extremely dangerous, and temperatures as low as 25°C are fatal. The temperature of the peripheral body tissues, particularly the skin, can safely vary over a much wider range. From a thermal point of view, the body can be considered to have a warm core where much of the heat is produced. This core is surrounded by a shell of cooler, insulating tissues (particularly subcutaneous fat). Under normal conditions, the difference between core temperature and skin temperature is about 4°C, but the difference may be as great as 20°C under severe climatic conditions (Sanders and McCormick 1993).

Core temperature can be maintained only if there is a state of heat balance with the environment. Even under comfortable climatic conditions, heat must be dissipated; otherwise, it will continue to build up and, eventually, lead to death. Under cool conditions, the body tries to limit the amount of heat that it loses to the environment. Little heat transfer with the environment takes place via conduction. Conduction is the process whereby heat energy is transferred from a warmer to a cooler object with which it has direct physical contact. The body tissues are poor heat conductors. Convection is the primary means through which heat is transferred to the skin from the deep tissues. Blood is an ideal medium of heat convection because its specific heat capacity and thermal conductivity are high.

Heat may also be gained from the environment or lost to the environment by convection or radiation. If the skin of a resting person is warmer than the surrounding air, the air at the skin surface is warmed by the conduction of heat from the skin. Since warm air rises, a flow of air around the person is established and heat is convected away from the person's body. Cool air moves closer to the skin to replace the rising warm air. The airflow is known as convection current. In practice, a resting person has a layer of slowly moving warm air at the surface of the skin. Movement of one's body or of wind can disrupt this warm layer and increase the rate of heat loss. Depending on the temperature of the air, this heat loss will be perceived as a chilly draft or a cooling breeze.

All objects at temperatures greater than absolute zero emit infrared radiation, which is sometimes referred to as radiant heat. The human body is no exception. If its temperature is greater than that of its surroundings, a radiant heat loss takes place. If the surroundings are hotter than the body, it experiences a net heat gain. Inside buildings, the walls and ceilings are common sources of radiant heat, whereas the sun is the major source outdoors. In the steel and glass manufacturing industries, furnaces are a major source of radiant heat.

Heat is lost to the environment by sweat production and evaporation; the body's tissues are composed largely of water, which is lost to the environment by diffusion from the skin, the lungs, and the sweat glands. If the temperature of the surroundings is greater than that of the body, no heat can be lost by convection or radiation, so evaporative heat loss by sweating is essential to maintain thermal balance.

In a cold environment, a state of thermal balance can be achieved by either increasing the rate of metabolic heat production (by shivering or by carrying out some fairly strenuous activity) or reducing the heat loss by convection and radiation (by wearing more clothes). In hot conditions, there is a high rate of metabolic heat production as well as heat gain from the environment (by radiation). A state of thermal balance can be obtained by either reducing the rate of metabolic heat production (by reducing the work rate or introducing rest pauses) or protecting the worker from the radiant heat (by means of a protective suit, for example).

5.2.1 Thermoregulatory Mechanisms

A number of physiological mechanisms exist for maintaining heat balance in both hot and cold environments.

5.2.1.1 Heat Stress

Under conditions of heat stress, the body absorbs and/or produces more heat than it can dissipate, leading to an increase in core temperature. The rise in core temperature results in an increase in metabolism. The increased metabolism generates more heat that must be dissipated; otherwise, core temperatures will increase even further, resulting in a further increase in metabolism and heat production that can lead to death. In many parts of the world, outdoor temperatures during summer can produce heat stress in individuals engaged in moderate physical activity. Jensen (1983) indicated that the highest rates of reported workers' compensation cases for heat illness were in agriculture (9.16 cases per 100,000 workers), construction 6.36 cases per 100,000 workers), and mining (5.01 cases per 100,000 workers). Another industry associated with heat illness is the iron and steel industry (Dinman *et al.* 1974). Heat stress can also occur in moderate or even cold climates where impermeable protective garments are worn and heavy physical activity is required. Examples include firefighters and toxic waste disposal workers. The National Institute of Occupational Safety and Health (NIOSH 1986) conservatively estimates that 5–10 million workers in the United States work in industries where heat stress is a potential safety and health hazard.

The first line of defense against heat stress is the cardiovascular system. Two basic responses are involved. Firstly, in hot environments, peripheral vasodilation occurs in the skin circulation. The arterioles dilate and capillaries at the skin surface open, blood flow increases, and heat is conducted to the skin surface over a small distance, from where it is dissipated to the environment. If the temperature of the air and surroundings is lower than that of the skin, heat is lost through convection and radiation. When in a comfortable environment, the skin blood flow amounts to about 5% of the cardiac output. In extreme heat, it can increase to 20% or more (Astrand and Rodahl 1977).

Secondly, in response to heat stress, there is an increase in heart rate. Under heat stress, cardiac output can increase 50%–75%, with the excess output diverted to the skin for cooling purposes. The need to dissipate heat in hot environments creates competition for blood between the skin and the working muscles. There may even be insufficient oxygen delivered to working muscles during heat stress thus causing lactic acid to build up and muscle pain and fatigue to occur. Recovery from physical work also takes longer in hot environments because blood is needed to dissipate heat and remove lactic acid.

The body's second line of defense against heat stress is increased sweating. Heat is lost by evaporation of the sweat. Sweat that drips off the body rather than being evaporated results in virtually no heat loss from the body; it is just wasted water loss. Excessive sweating can cause dehydration if fluids are not replaced by drinking. Dehydration itself affects thermoregulation and results in a rise in core temperature. People working in a hot environment must be encouraged to drink in order to replace lost fluid to avoid dehydration. Water should be taken at frequent intervals. In hot environments, NIOSH (1986) recommends drinking 150–200 ml of water every 15–20 min.

5.2.1.2 Cold Stress

With respect to severity, cold stress is less of an occupational health hazard than heat stress. Although one can die from cold stress, few people do, whereas death from heat stroke is more common. There are only a few circumstances in which people are required to work in cold environments. They include (a) outdoor work in winter, (b) work in arctic locations (especially in the military and oil and gas exploration), and (c) work in cold-storage warehouses. A decrease in core temperature is accompanied by a decrease in the metabolic rate and, therefore, the amount of metabolic heat production. If metabolic heat production cannot be elevated while heat is being lost, a viscious circle of further heat loss and still lower metabolic heat production may develop.

Vasoconstriction and shivering are the two primary physiological reactions to cold stress. Each is aimed at reducing the net heat loss from the body in order to maintain a constant internal temperature in a cold environment. The body's first defense against cold stress is to constrict the blood vessels in the skin and extremities (i.e. vasoconstriction). Blood flow to the skin and extremities can be virtually shut off to these regions. This serves two purposes. First, the warm blood is kept away from the cold skin so that less heat is lost to the environment. Second, the insulating capacity of the skin is increased up to sixfold by cutting off the flow of blood (Sanders and McCormick 1993). Because of this, the temperature of the fingers and toes can rapidly approach the temperature of the surrounding air, causing local cold injury (e.g. frostbite). Vasoconstriction also causes more blood to be distributed to the internal organs. Consequently, more blood is passed through the kidneys, which causes an increased production of urine. This is why people urinate more frequently in cold environments than in warm environments. Vasoconstriction also deprives the smooth muscles that constrict the blood vessels of oxygen. Eventually they fatigue and allow the blood vessels to dilate. The blood that flows in has not been circulating and has become deoxygenated. This deoxygenated blood gives the skin the characteristic blue appearance that is associated with prolonged exposure to cold stress.

While vasoconstriction occurs in most areas of the body, the head is a noteworthy exception. Up to 25% of all heat loss can take place at the head (Kroemer 1991). For this reason, the head should be considered an extremity and protected accordingly, much like the hands and feet.

Involuntary shivering is a thermoregulatory mechanism involving active heat production. Groups of motor units act out of phase with one another, and muscles act against their antagonists (Bridger 1995). Almost no movement occurs, and the result is a high level of heat production. In fact, the increased muscular activity increases metabolic heat from two to four times what it was at thermal equilibrium at rest (Sanders and McCormick 1993). This can be enough to offset the loss of heat to the environment and bring the net heat storage to zero. However, it can never be enough to turn the heat storage positive. That is, shivering will never heat a person up; it can only stop him/her from cooling down.

Voluntary movement also increases heat production but breaks up the insulating layers of air around the body, which increases the rate of heat loss. Both shivering and voluntary movement increase oxygen uptake and cardiac output and can lower a person's capacity to carry out physical work. It then follows that the more fit a person is, the more efficiently he/she can produce heat by shivering and the longer the shivering reflex can be maintained without suffering exhaustion. Good physical conditioning can mean the difference for surviving severe cold stress.

5.3 WORK IN HOT CLIMATES

Heavy physical work in the heat imposes conflicting demands on the cardiovascular system. Peripheral vasodilation requires an increase in blood flow to the skin (up to 10 l per minute). However, working muscle also demands increased blood supply compared with relaxed muscle (up to 25 l per minute). Since cardiac output cannot exceed venous return and the average maximum cardiac output is about 25 l per minute, cardiac capacity is a limiting factor for muscular work in the heat (Bridger 1995). The cardiovascular system may be placed under considerable strain when a person is working in the heat, as output rises to meet the demands of both physical work and bodily cooling. A dangerous condition can arise if the heart is no longer able to meet both demands.

5.3.1 Heat Illness

Under extreme or prolonged heat stress, several disorders can result. The most severe, heat stroke, can lead to death. The remainder of this section describes several forms of heat illness.

5.3.1.1 Heat Exhaustion

Lamb (1984) refers to heat exhaustion as heat syncope. This illness is characterized by muscular weakness, nausea, vomiting, dizziness, and fainting (syncope). The principal cause is dehydration and is more likely to affect individuals who are not heat acclimatized and are in poor physical condition.

5.3.1.2 Heat Cramps

There may be painful spasms of muscles used during work (notably in the arms, legs, and abdomen). The cramps occur during or hours after working in a hot environment. Heat cramps are associated with restricted salt intake and profuse sweating, causing excessive loss of salt. This condition may be alleviated by giving the victim commercial electrolyte drinks or several glasses of water, each containing a half-teaspoon of dissolved table salt (Lamb 1984).

5.3.1.3 Heat Stroke

Heat stroke (thermoplegia) is a dangerous condition caused by exposure to excessive heat. Sweating stops, and the skin becomes dry. The victim experiences dizziness, headache, nausea, muscle cramps, and a dangerously high core temperature. Since the core temperature control system is located in the brain, the elevated temperature may disrupt the thermoregulatory system itself, and even if the worker stops working, core temperature may not return to normal. Rapid cooling of the body using cold sprays or wet sponges can be used to lower the temperature.

5.3.2 Heat Acclimatization

Heat acclimatization is a physiological process of adaptation rather than a psychological adjustment to life in a hot environment. It involves an increase in the capacity to produce sweat and a decrease in the core temperature threshold value for the initiation of sweating. The maximum rate of sweat production can double from 1 liter per hour in an unacclimatized person to 2 l per hour in an acclimatized person (Bridger 1995).

Sanders and McCormick (1993) state that most of the acclimatization changes have taken place within 4–7 days of exposure to a hot environment. At the end of 12–14 days, the process is complete. Acclimatization also occurs more quickly if one is physically fit (Lamb 1984). Vitamin C (ascorbic acid) may also enhance heat acclimatization (Strydom *et al.* 1976). For acclimatization, NIOSH (1986) recommends specific schedules for new workers and those returning to the hot environment after a prolonged absence.

Although heat acclimatization occurs naturally, it may also be induced artificially. In some mines, workers are artificially acclimatized to heat as a requirement of their employment (Bridger 1995). Surface acclimatization chambers are used, and workers exercise (by performing step-ups) in hot conditions. Other methods of acclimatization include exercise while wearing vapor barrier suits (which prevent sweat evaporation) and resting in baths at 41°C.

An individual's ability to produce sweat depends on the climate experienced in the early years of life (Diamond 1991). It appears that the sweat reflex develops in the first 2.5 years of life. Although we are all born with a similar number of sweat glands in the skin, the number that develop, or are "switched on," depends on the conditions experienced early in life. People who grew up in a hot environment develop a larger number of functional sweat glands than those who grew up in a cold environment and can therefore tolerate hot conditions more easily.

There is wide variability among people in terms of their ability to acclimatize to heat. Therefore, even with a regimented program of heat acclimatization, individuals working in hot climates must be physiologically monitored during acclimatization to assess their level of physiological strain.

Although the body can acclimatize to heat, it cannot acclimatize to dehydration. Fluid must be made available at all times, even for acclimatized workers.

5.3.3 Heat Stress Management

Sanders and McCormick (1993) suggest some approaches for the elimination of heat stress. They include the following:

- Reduce high relative humidity by using dehumidifiers.
- Increase air movement by using fans or air conditioners.
- Suggest that workers remove heavy clothing; issue loose-fitting overalls.
- Reduce the work rate.
- Include frequent rest pauses.
- Introduce job rotation.
- Carry out outdoor work at cooler times of the day (e.g. early morning).
- Allow two weeks for acclimatization.
- Enforce rest breaks and provide drinking water or other fluids.
- Provide shade to reduce radiant heat load (plant trees, issue wide-brimmed hats, etc.).
- In factories, build cool spots and refuges to lower worker exposure.

5.4 WORK IN COLD CLIMATES

Core temperatures can be maintained in the cold if the person is working and suitable protective clothing is provided. When manual work is not being performed, less metabolic heat is generated, and adequate insulation becomes of increasing importance. Metabolic heat production may increase by a factor of three when a person is working compared with resting (Bridger 1995). When the person is resting, therefore, three times the amount of clothing insulation is required. Counter-intuitively, adequate hydration is essential when one is working in a cold environment because sweating may still take place.

5.4.1 Severe Cold Stress

The most common cold injury is frostbite. Frostbite is the freezing of body tissue and the formation of ice crystals in the tissue cells. Frostbite can result from local cooling, most commonly to the hands and feet. Even if core temperature is maintained at normal levels, contact with metal or liquid (good conductors of heat) can cause the skin to freeze. The real danger comes when freezing occurs in the deeper tissue (second- and third-degree frostbite). The ice crystals can rupture the tissue cells, and blood cells can aggregate in the vessels, leading to gangrene.

Hypothermia is generalized cold stress and is usually characterized by a body temperature below 35°C. The initial stages are marked by disorientation, apathy, hallucinations, aggression or euphoria. As body temperature drops further, the person may become unconscious, cardiac arrhythmia may develop, the pupils of the eyes may become dilated, and finally, the heart may stop. In extreme cold stress, the body's rate of metabolism slows, and the body (the brain included) needs less oxygen. This, in part, explains the miraculous survival of people who drown in frozen lakes and, even after 10 minutes under water, are revived with little or no brain damage. At normal body temperatures, three minutes without oxygen would mean almost certain brain damage.

5.4.2 Acclimatization to Cold

There is some scant physiological evidence to suggest that, in humans, local acclimatization to cold may occur in the extremities. Increased blood flow through the hands may occur after repeated exposure to cold conditions. This preserves manual dexterity but at the cost of increased heat loss from the body as a whole. Fish filleters, for example, who habitually expose their hands to cold water have an increased blood flow through their hands so that their hands remain warm and are not so apt to becoming numb when exposed to the cold water (Nelms and Soper, 1962).

Behavioral adaptation, as compared to physiological acclimatization, may play a larger role in a human's ability to work in cold climates. Through experience, they wear proper clothing. Clothing provides a subtropical microclimate of warm, moist air between the skin and clothes. Most people put on more clothes in cold conditions in order to maintain this microclimate and, for this reason, are rarely exposed to any stimulus for true acclimatization to the cold. People also increase their daily food intake, which serves to increase their metabolic rate and the thickness of the subcutaneous fat layer, resulting in a higher rate of heat production and better insulation of the core tissues. The chemical processes involved in the digestion of food are also a source of heat.

REFERENCES

Astrand, P. O., & Rodahl, K. (1977). *Textbook of work physiology*. New York: McGraw-Hill.

Bridger, R. S. (1995). *Introduction to ergonomics (Chapter 9)*. New York: McGraw-Hill.

Diamond, J. (1991). Pearl Harbor and the emperor's physiologists. *Natural History*, *12*, 3–7.

Dinman, B., Stephenson, R., Horvath, S., & Colwell, M. (1974). Work in hot environments: I. Field studies of workload, thermal stress, and physiologic response. *Journal of Occupational Medicine*, *16*(10), 785–791.

Jensen, R. (1983). Workers' compensation claims related to heat and cold exposure. *Professional Safety*, *28*(9), 19–24.

Kroemer, K. H. E. (1991). Working strenuously in heat, cold, polluted air, and at altitude. *Applied Ergonomics*, *22*(6), 385–389.

Lamb, D. R. (1984). *Physiology of exercise: Responses and adaptations* (2nd ed.). New York: Macmillan.

National Institute of Occupational Safety and Health. (1986). *Criteria for a recommended standard: Occupational exposure to hot environments, revised criteria 1986*. Washington, DC: Superintendent of Documents.

Nelms, J., & Soper, J. (1962). Cold vasodilation and cold acclimatization in the hands of British fish filleters. *Journal of Applied Physiology*, *49*(4), 444.

Sanders, M. S., & McCormick, E. J. (1993). *Human factors in engineering and design* (7th ed., Chapter 17). New York: McGraw-Hill.

Strydom, N. B., Kotze, H. F., Van Der Walt, W. H., & Rodgers, G. G. (1976). Effect of ascorbic acid on the rate of heat acclimatization. *Journal of Applied Physiology*, *41*(2), 202–205.

FURTHER READING

Cibis, T., & Carolyn McGregor, A. M. (Eds.). (2022). *Engineering and medicine in extreme environments*. Springer.

Guo, H., & Ofori, A. A. (2021). The internet of things in extreme environments using low-power long-range near field communication. *IEEE Internet of Things Magazine*, *4*(1), 34–38.

Mondal, K., Nuñez III, L., Downey, C. M., & Van Rooyen, I. J. (2021). Recent advances in the thermal barrier coatings for extreme environments. *Materials Science for Energy Technologies*, *4*, 208–210.

Richardson, E. E. (2024). Predicting and Protecting the Cognitive Health of Operators in Isolated, Confined, and Extreme Environments. In *Proceedings of the 23rd International Conference on Autonomous Agents and Multiagent Systems* (pp. 2764–2766). Auckland, New Zealand.

Suedfeld, P. (2022). Adventures in extreme environments. In *Adventure psychology* (pp. 42–59). Routledge.

Work in Extreme Environments: Effects on Performance

6

M. Kolich

6.1 INTRODUCTION

Environmental conditions can have a profound effect on work performance. To this end, the National Institute for Occupational Safety and Health (NIOSH) has established upper limits for occupational exposure (NIOSH 1986). This chapter, while building on the physiological fundamentals presented in part I, presents a summary of the performance effects associated with work in extreme environments.

There are two main theories concerning the effect of the environment on performance. Arousal theory states that, for all tasks, there is an optimum level of arousal (readiness to act) at which maximum performance occurs. Environmental extremes can increase the arousal level, whereas overly comfortable conditions can lower it. If the arousal level deviates significantly from the optimum, performance deteriorates. A competing theory suggests that environmental extremes, particularly cold, have a distracting effect—performance declines

DOI: 10.1201/9781003512257-6

because of momentary shifts of attention away from the task toward the environment. The remainder of this chapter highlights the pertinent research.

6.2 EFFECTS OF HEAT ON PERFORMANCE

The results of studies investigating the effects of heat stress on performance have been somewhat contradictory. For example, some studies have reported little, if any, performance loss under extreme temperature exposure, whereas other studies have reported performance decrements. Some of this is undoubtedly due to differences among studies with regard to exposure conditions, the motivation, acclimatization, and skill level of the subjects used, and the nature of the tasks performed. Despite this state of affairs, a few conclusions and reasonable hypotheses can be extracted from the studies and reviews of the literature.

6.2.1 Physical Work

Heat stress creates competition for blood between the working muscles and the skin. Exhaustion from heavy physical work therefore occurs much sooner in a hot environment than in a temperate one. For example, people sitting at rest can tolerate 38°C for three or more hours. Working at 280 kcal/hr, however, reduces tolerance time to about 30 minutes under such conditions (Sanders and McCormick 1993).

6.2.2 Simple Cognitive and Perceptual–Motor Performance

There have been several reviews of the effects of heat on cognitive and perceptual–motor performance. In most of these studies it was found that the type of task mediates the effects of heat stress on performance. Several taxonomies of tasks have been suggested. A simple and useful taxonomy divides cognitive and perceptual–motor tasks into two categories: simple and complex. Simple tasks include visual and auditory reaction time, arithmetic problem solving, coding, and short-term memory tasks. Performance decrements on simple cognitive and perceptual–motor tasks are commonly not observed in the heat, and in fact, an enhancement in performance is frequently observed during brief exposures for these types of tasks. This contention is in accord with the arousal theory previously mentioned.

Tracking and vigilance tasks can be considered complex. Astrand and Rodahl (1977) indicate that rather consistent performance decrements start at about 30°C–35°C wet-bulb globe temperature (WBGT). Wyon (1974) investigated the effects of moderate heat stress on typewriting performance by reanalyzing the data from a 1923 report by the New York State Commission on Ventilation and showed that people consistently produce more at 20°C than 24°C. Therefore, it seems that mild heat stress can cause drops in performance. More severe heat stress may incapacitate workers as well as over-arouse them.

Azer *et al.* (1972) provide more evidence that hot conditions have significant effects on performance. In their experiment, performance decrements occurred when subjects worked at 35°C and 75% relative humidity, but not at 35°C or 37.5°C and 50% relative humidity. Relative humidity seems to be a key determinant of performance in a variety of tasks. Allen and Fischer (1978) found that if relative humidity was held constant at 40%, performance on a simple learning task did not vary over the range of 11°C–28°C. If relative humidity was not controlled, optimum performance was found to occur at 18°C, 35% relative humidity. Increases and decreases in temperature lead to performance decrements.

More recently, Vasmatzidis *et al.* (2002) investigated the adequacy of the NIOSH recommended heat stress standard in terms of protecting time-sharing performance. They concluded that the NIOSH criterion does offer protection up to 28°C WBGT.

6.2.3 Safety Behavior

Ramsey *et al.* (1983) observed workers' behaviors over a 14-month period under varying climatic conditions. Over 17,000 observations were made. The ratio of observed unsafe behaviors to the total number of observed behaviors formed a U-shaped curve when plotted over ambient temperature. The ratio of unsafe behaviors was at a minimum between 17°C and 23°C WBGT. When climatic conditions were below or above this generally preferred range, the incidence of unsafe behaviors increased. This is further support for the value of providing people with a comfortable climate in which to work.

Another study that has implications for safe behavior was conducted by Colquhoun and Goldman (1972). These researchers investigated the ability of subjects to detect target stimuli against a noisy background at a temperature of 38°C dry-bulb, 33°C wet-bulb. It was found that only when work in the heat was accompanied by an actual increase in body temperature was an effect on performance observed. The effect was to increase subjects' confidence by saying that they had detected a target rather than to increase the detection rate. In fact, subjects made more false positive errors (saying that they had detected a target when there was none). Elevated body temperature may lead to more risky behavior. This research has relevance for quality control and inspection performance in all "hot" industries, such as iron, steel and glassmaking.

6.2.4 Individual Differences Affecting Performance in Hot Conditions

There are considerable differences between people with respect to their tolerance to heat stress, and therefore, performance in hot conditions. This section highlights several factors that account for some of these differences.

6.2.4.1 Age

The very young and the very old are less tolerant than other age groups. Young children have less sweating capacity than adults. Older men are less able to tolerate high heat stress partly because of their higher skin temperature threshold for the onset of sweating (Bridger 1995). In addition, aging results in a more sluggish response of the sweat glands and less total body water (Sanders and McCormick 1993). This results in less effective thermal regulation and compromises performance.

6.2.4.2 Gender

Under heat stress, women generally show greater increases in heart rate, lower maximal sweat rates, higher skin temperatures, and, in severe wet heat, higher core temperatures. They are affected more by dehydration, and fewer women than men can be successfully heat-acclimatized. A major factor accounting for these results is differences in physical fitness between men and women, although physically fit women respond to heat stress in a manner similar to men of equal fitness. Women have a higher proportion of body fat than men, which may also play a role (see below).

6.2.4.3 Body Fat

Excess body fat degrades heat tolerance by increasing the mass-to-surface area ratio of the body and reducing cardiovascular fitness. Also, adipose (fat) tissue contains less water than other tissues and has a lower specific heat capacity (Bridger 1995). Thus, the same heat load will cause a greater temperature increase in obese individuals compared with lean individuals.

6.2.4.4 Physical Fitness

Physical fitness improves heat tolerance because both rely on cardiovascular function and sweat production. Fit individuals can perform a given level of physical work with smaller increases in heart rate and less heat production than unfit individuals. This is due in part to the greater capacity of fit individuals to deliver oxygen to the working muscles. The fit individual can then divert more blood to the skin for cooling without sacrificing the oxygen needs of the muscles (Sanders and McCormick 1993). For this reason, physically fit workers are less stressed by hot conditions even if they are accustomed to a temperate climate.

6.2.4.5 Alcohol

Alcohol has been commonly associated with heat stroke. It interferes with central and peripheral nervous functioning and is associated with dehydration. Ingestion of alcohol prior to or during work reduces heat tolerance and increases the risk of heat stroke (Sanders and McCormick 1993).

6.3 EFFECTS OF COLD ON PERFORMANCE

The effects of cold on the performance of various types of human functions are not yet fully understood and, in any event, are rather intricate. Therefore, it is often difficult to disentangle the effects of cold on the psychological and physical aspects of task performance. Cold causes peripheral vasoconstriction, shivering, impaired neural conduction, and poorer muscular control—all of which may have effects, such as longer reaction time, which may appear to be of psychological nature. Some of the interacting factors that make the study of the effects of cold on performance complicated include the specific type of task or function, the interaction of air temperature, humidity, air flow, and radiation, duration of exposure, whether the body is being warmed or cooled, differential exposure to different parts of the anatomy (e.g. body versus hands), acclimatization and individual differences. Given this assortment of interacting variables, it is still possible to distil at least a few generalizations from the available research. A few such generalizations are given in this section.

6.3.1 Physical Work

Cold has a detrimental effect on physical work. A reduction in core temperature and/or muscle temperature reduces physical functions, resulting in a decrease in muscle strength and endurance (Sanders and McCormick 1993). A couple of factors account, at least in part, for this reduced physical capacity in the cold. For one, reduced core temperature reduces the rate of metabolism within the muscle. For another, cold reduces the speed of neural conduction in the peripheral motor nerves (Sanders and McCormick 1993).

6.3.2 Tactile Sensitivity

Tactile sensitivity is related to finger skin temperature. Enander (1984) believes that it is actually the more slowly changing temperature of the deeper tissue rather than the actual surface temperature that negatively affects sensitivity. In most cases, however, the two temperatures are closely related (Sanders and McCormick 1993). This reduction in sensitivity makes the performance of tasks that require manipulation of small objects difficult. Assembly and repair tasks, for example, are adversely affected by the cold.

6.3.3 Manual Performance

Related to tactile sensitivity, manual performance is, without doubt, adversely affected by the cold. Manual performance seems to be negatively related to hand–skin temperature (Sanders and McCormick 1993). In addition, Lockhart *et al.* (1975) found that slow cooling results in a greater performance decrement than fast cooling at equivalent hand-skin temperatures. This indicates that it is the temperature of the deeper tissue that is responsible for the decline in performance.

6.3.4 Tracking Performance

Tracking performance is adversely affected by the cold. Although the lower limit of unimpaired performance is still uncertain, it is probably between an ambient temperature of 4 and 13°C (Sanders and McCormick 1993). At low temperatures there is often a marked loss of motivation and increase in apathy among subjects (Payne 1959), which may account, in part, for the decrement in performance.

6.3.5 Reaction Time

There is no systematic evidence that simple reaction time is affected by cold. There is, however, evidence that choice reaction time is adversely affected. What is typically found is that cold increases the number of errors made on the task but decreases the reaction time on trials in which errors are made. Reaction time on correct trials is not affected.

6.3.6 Mental Activities

Evidence about the effects of cold on complex mental activities is ambiguous. Some investigators have found virtually no effect, while others have found some decrement. Sanders and McCormick (1993) suggest that cold may act as a distracter that interferes with some types of mental performance. The effect would, of course, depend on the type of task, the severity of the cold, the skill level of the subjects, and prior experience of the subjects with performing in the cold.

6.3.7 Individual Differences Affecting Perception of Cold

The perception of cold seems to depend on experience but it can affect performance. For example, workers accustomed to a cold climate may feel comfortable when their deep body tissues are suitably insulated by clothing, despite local cooling of the skin of the cheeks, ears, nose, fingers, and feet. Cold perception also depends on the "set point" of the hypothalamic temperature-regulating center. Bacterial infections (which cause chills, for example) can elevate the set point for core temperature thermoregulation. The person, therefore, feels cold even though the core temperature is normal. Peripheral vasoconstriction takes place, and the person shivers. After several hours, the core temperature reaches the new set point, and the person again feels neither hot nor cold. When the factor that initially caused the increase in set point is removed, the person feels hot, peripheral vasodilation and profuse sweating occur as the core temperature declines and reaches the original set point.

6.4 EFFECTS OF DRAUGHT ON PERFORMANCE

Draught is the most annoying climatic factor in many workplaces. Its sources are leaky windows and open doors, or cooling devices where the outlets are installed in the ceiling. Draught is characterized by a varying air velocity, which is quantified by turbulence intensity (measured as the standard deviation of air speed divided by mean air speed. Griefahn *et al.* (2002) produced a model to suggest that annoyance increased with the increase in mean air velocity and in turbulence intensity, as well as with the decrease in air temperature and in metabolic rate. The model was proposed as a substitute for the model in ISO 7730 (1994).

6.5 EFFECTS OF ENVIRONMENTAL CONDITIONS ON THE MUSCULOSKELETAL SYSTEM

Hildebrandt *et al.* (2002) performed an epidemiological study to assess the generation-old observation (i.e. “old wives tale”) that environment influences musculoskeletal aches. They analyzed questionnaire data collected on musculoskeletal symptoms, workload, and perceptions of climatic conditions for 2030 workers in 24 different occupations. The results revealed that approximately 25% of the workers related symptoms of the low back and neck–shoulders to climatic factors, particularly draughts and wind. The strength of this finding warrants further investigation.

6.6 CONCLUSION

The main conclusion to be drawn from this particular chapter is that human beings have a remarkable capacity to adapt. That is, a Canadian gets along better in the cold than does a Mexican for the same reason as a Mexican gets along better in heat than does a Canadian. Simply put, it is a matter of the individual getting used to and adapting to a way of life compatible with the prevailing conditions of the area in question. Industrial ergonomists have a role to play in facilitating this adaptation.

REFERENCES

Allen, A. A., & Fischer, G. J. (1978). Ambient temperature effects on paired associate learning. *Ergonomics, 21*(2), 95–101.

Astrand, P. O., & Rodahl, K. (1977). *Textbook of work physiology.* New York: McGraw-Hill.

Azer, N. Z., McNall, P. E., & Leung, H. C. (1972). Effects of heat stress on performance. *Ergonomics, 15*(6), 681–691.

Bridger, R. S. (1995). *Introduction to ergonomics (Chapter 9).* New York: McGraw-Hill.

Colquhoun, W. P., & Goldman, R. F. (1972). Vigilance under induced hypothermia. *Ergonomics, 15*(6), 621–632.

Enander, A. (1984). Performance and sensory aspects of work in cold conditions: A review. *Ergonomics, 27*(4), 365–378.

Griefhn, B., Kunemund, C., & Gehring, U. (2002). Evaluation of draught in the workplace. *Ergonomics, 45*(2), 124–135.

Hildebrandt, V. H., Bongers, P. M., Van Dijk, F. J. H., Kempers, H. C. G., & Dul, J. (2002). The influence of climatic factors on non-specific back and neck-shoulder disease. *Ergonomics, 45*(1), 32–48.

ISO 7730. (1994). *Moderate thermal environments: Determination of the PMV and PPD indices and specification of the conditions for thermal comfort.* Geneva: ISO.

Lockhart, J. M., Kiess, H. O., & Clegg, T. J. (1975). Effect of rate and level of lowered finger-surface temperature on manual performance. *Journal of Applied Psychology, 60*(1), 106–113.

National Institute of Occupational Safety and Health. (1986). *Criteria for a recommended standard: Occupational exposure to hot environments, Revised criteria 1986.* Washington, DC. Superintendent of Documents.

Payne, R. (1959). Tracking performance as a function of thermal balance. *Journal of Applied Physiology, 14*, 387–389.

Ramsey, J., Burford, C., Beshir, M., & Jensen, R. (1983). Effects of workplace thermal conditions on safe work behaviour. *Journal of Safety Research*, *14*, 104–114.
Sanders, M. S., & McCormick, E. J. (1993). *Human factors in engineering and design* (7th ed., Chapter 17). New York: McGraw-Hill.
Vasmatzidis, I., Schlegel, R. E., & Hancock, P. (2002). An investigation of heat stress effects on time-sharing performance. *Ergonomics*, *45*(3), 218–239.
Wyon, D. P. (1974). The effects of moderate heat stress on typewriting performance. *Ergonomics*, *17*(3), 309–318.

FURTHER READING

Habibi, P., Moradi, G., Dehghan, H., Moradi, A., & Heydari, A. (2021). The impacts of climate change on occupational heat strain in outdoor workers: A systematic review. *Urban Climate*, *36*, 100770.
Ioannou, L. G., Foster, J., Morris, N. B., Piil, J. F., Havenith, G., Mekjavic, I. B., ... & Flouris, A. D. (2022). Occupational heat strain in outdoor workers: A comprehensive review and meta-analysis. *Temperature*, *9*(1), 67–102.
Jay, O., Capon, A., Berry, P., Broderick, C., de Dear, R., Havenith, G., ... & Ebi, K. L. (2021). Reducing the health effects of hot weather and heat extremes: From personal cooling strategies to green cities. *The Lancet*, *398*(10301), 709–724.
Périard, J. D., Eijsvogels, T. M., & Daanen, H. A. (2021). Exercise under heat stress: Thermoregulation, hydration, performance implications, and mitigation strategies. *Physiological Reviews*. https://doi.org/10.1152/physrev.00038.2020
Wolkoff, P., Azuma, K., & Carrer, P. (2021). Health, work performance, and risk of infection in office-like environments: The role of indoor temperature, air humidity, and ventilation. *International Journal of Hygiene and Environmental Health*, *233*, 113709.

7 Mental Workload under Thermal Stress

B. Chase

7.1 INTRODUCTION

Human performance of cognitive tasks is affected by a variety of psychological, physiological, and environmental variables. Exposure to either excessive cold or heat can produce performance decrements on simple cognitive tasks and on more complex perceptual motor tasks. The majority of research in this area has focused on cognitive performance in heat, and this will also be the focus here.

Mental workload can be thought of as the amount of mental effort required for task performance. The assessment of mental workload is a process of applying models of human cognition to the analysis of tasks and has as its goal maintaining mental workload at a level that allows acceptable task performance.

For over 50 years researchers have been examining the effects of thermal stress on human cognitive performance. The research thus far paints a somewhat fuzzy picture of the effects of thermal stress on mental workload.

DOI: 10.1201/9781003512257-7

7.2 THERMAL ENVIRONMENT

7.2.1 Measurement

Past studies of human performance under thermal stress have employed a variety of measurement techniques. Many studies either did not specify or did not fully specify the environmental conditions. Recent research employs wet bulb globe temperature (WBGT).

7.2.2 Dry-Bulb Temperature

Most people are familiar with standard household thermometers. Dry bulb temperature is simply the reading of a typical mercury thermometer. The result is a measure of the temperature of the surrounding air. However, there are often other factors that can affect the heat stress of a human.

7.2.3 Wet-Bulb Temperature

Wet-bulb temperature is obtained by placing a wet wick over the mercury bulb, and when air passes over the wick, evaporation and consequently cooling occur. The rate of cooling from evaporation is dependent upon the humidity of the surrounding air. Consequently, the wet- and dry-bulb temperatures can be used to calculate the relative humidity.

7.2.4 Globe Temperature and WBGT

A well-known index is WBGT, which attempts to combine the effects of air temperature, water vapor pressure, air speed, and radiant temperature into a single, robust value. WBGT is calculated using:

$$\text{WBGT}\left(\text{outdoor conditions}\right) = 0.7\,\text{wbt} + 0.2\,\text{gt} + 0.1\,\text{db}$$
$$\text{WBGT}\left(\text{indoor conditions}\right) = 0.7\,\text{wbt} + 0.3\,\text{gt},$$

where wbt is wet bulb temperature, dbt is dry bulb temperature, and gt is globe temperature. Globe temperature is obtained by taking a thin copper sphere, painted flat black, and placing a thermometer at the center, and allowing the thermometer to reach equilibrium.

7.3 ATTENTION AND COGNITIVE PERFORMANCE

7.3.1 Attention

The general topic of attention has been investigated since the turn of the century. Scientific inquiry into the subject of attention, while popular around the turn of the century, lost recognition during the first half of the twentieth century (possibly due to the influence of Behaviorism). During modern applied research, attention was of interest as the devices humans had to operate became more complex and consequently more demanding. Of interest was the ability to attend to many objects of stimuli at once, such as in guiding a land, air, or sea vessel. These activities involve elements of attention like tracking, vigilance, decision-making, communicating, etc.

Fundamental aspects of attention can be described, such as selective attention, or the ability to concentrate on certain elements of one's environment while ignoring others. Divided attention describes the ability to divide attention among various tasks, which may be competing for our attentional resources. The idea of *mental workload* is meant as a measure or assessment of the attentional demands a task may place on a human operator.

7.3.2 Models of Attention

There is a variety of competing models in the area of attention. The current available models can be divided into two types as they pertain to human factors. The first type is often termed a bottleneck model, which refers to a point in the process where a limit of attention may occur. This type of model may be contrasted with a resource model, where attention is viewed as a capacity that has limits. Further, these two models may each be divided into categories. Bottleneck models can be divided into early selection and late selection. Resource models can be divided into single resource and multiple-resource theories.

7.3.3 Bottleneck Models

In 1958 Broadbent proposed a model known as filter theory. In filter theory, an early selection model, stimuli enter the central processor individually and are filtered according to whether they are wanted or extraneous items. Later, other models were proposed in which the filter did not serve to block unattended stimuli, but rather attenuated their signal. Others argued for a late selection model, which would place the screening process later in the sequence of events. The main impetus for this approach was the belief that even unwanted stimuli were at some point being identified. While the issue is quite difficult, in an attempt to deal with conflicting evidence, some researchers felt that the bottleneck was not fixed. Rather there could be a shift from early to late selection in the attentional process.

7.3.3.1 Resource Models

One of the main proponents of the resource model was Kahneman (1973). In his model, attention is viewed as a unitary, limited capacity resource. Owing to difficulties in experimental findings, some were led to propose a multiple resource model, which holds that there is no single attentional resource. Rather, there are multiple resources from which we may draw, which may or may not overlap (Proctor and van Zandt 1994).

Basic aspects of attention can be distinguished, such as *divided attention*. This refers to the ability to divide attention between two (dual) or more simultaneous tasks. Dual-task performance has been studied in great depth in the human factors/ergonomics literature. Dual tasks are more difficult and demand perceptual motor skills, more closely approximating industrial and military tasks (Ramsey 1995).

7.3.4 Attention and Arousal

The restriction of attention that occurs during high periods of arousal has been termed perceptual narrowing (Kahneman 1973). An individual's attentional capacity is influenced by his or her arousal. The level of arousal may exert an influence on the amount of attentional resources available. This relationship between arousal and performance underlies a fundamental law known as the Yerkes–Dodson Law (Yerkes and Dodson 1908). This inverted U-shaped relationship is illustrated in Figure 7.1.

The Yerkes–Dodson Law states that as arousal increases, performance initially increases. At the point where arousal, presumably from some stressor, exceeds the organism's ability to cope performance begins to deteriorate. Separate curves can be drawn for different classes of tasks. Simple cognitive tasks appear to be less sensitive to the level of arousal, whereas complex perceptual motor tasks seem to show a greater sensitivity to the level of arousal.

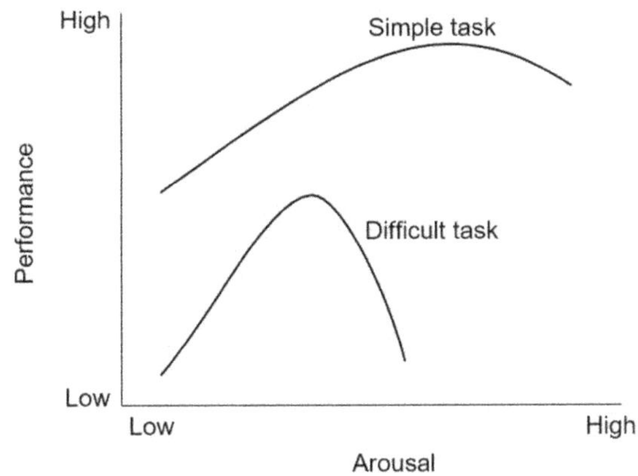

FIGURE 7.1 Yerkes–Dodson Law.

7.3.5 Dual Tasks

In situations where an operator's attention must be divided between two competing tasks, the tasks may not have the same priority or emphasis level. In cases such as these, performance on one task can be plotted along with performance on the second task.

The performance trade-off or dual-task cost can be captured graphically with a performance operating characteristic (POC) curve. POC is similar to the receiver operating characteristic (ROC) curve, which can plot hit rates against false-alarm rates in signal detection experiments. POC curves allow one to examine such aspects of performance as allocation of attention, divided attention, dual-task cost, and an operator's functional performance region. With these metrics, relative divided attention cost measures can be obtained and compared, which can control for individual differences in task performance.

7.4 MENTAL WORKLOAD METHODS

7.4.1 Measurement

There are multiple methods for the assessment of mental workload. The results of these assessments can prove useful in system design and redesign. Further, as the system is broken down into its constituent parts and the workload requirements of the system components analyzed, predictive models of performance can be constructed.

7.4.2 Assessment Techniques

7.4.2.1 Primary Task Measures

Measures of the primary system task can reflect the experimenter's interest, and overall system goals determine workload requirements and the choice of these measures on the task of interest.

7.4.2.2 Secondary Task Measures

It is assumed that performance on the primary task consumes a certain amount of cognitive resources, and a secondary task will use any resources left over from performance of the primary task. Consequently, if the primary task leaves few resources, performance will suffer on some secondary task, and performance on this secondary task becomes a measure of mental workload.

When using this method of task analysis, it is important to keep in mind the fact that the secondary task should not seem artificial and distracting.

7.5 PHYSIOLOGICAL MEASURES

Physiological measures can often remove subjective elements from the analysis. Heart rate analysis can prove quite useful in estimating mental workload. At lower levels of workload, the heart rate frequency can fluctuate, whereas under higher workload heart rate demonstrates greater constancy.

7.6 SUBJECTIVE MEASURES

On the more subjective side, an easy method for estimating workload is simply to ask the operator how difficult the task is. While a highly subjective measure, questionnaires can be constructed in a manner that allows for greater accuracy in rating. The rating scales should be anchored by thorough descriptions of the extreme and middle scale values and should attempt to analyze the task on multiple dimensions.

7.6.1 Heat Stress Effects

Studies on the effect of heat have generally yielded inconsistent results. This is partly due to the fact that different kinds of cognitive work are differentially sensitive to thermal stress. The majority of the work in heat and human performance has been concerned with sustained attention or vigilance tasks. However, when analyzing performance on various tasks, one must consider the type of task being performed.

A comprehensive study that took into account the type of task involved and converted, where possible, all temperature measures over to WBGT is Ramsey (1995). By dividing task types into two categories, Ramsey acknowledged the differentially sensitive nature of task types. The two categories were: mental, cognitive, very simple perceptual motor, sensory time estimation, reaction time, etc., other perceptual motor tasks, including tracking, vigilance, vehicle or machine operation, complex or dual tasks, etc.

Category 1 can be thought of as simple cognitive tasks and category 2 can be referred to as complex or dual tasks. Ramsey's conclusion in regard to the category 2 tasks (above) is that there is an onset of a statistically significant decrement in performance in the range 30–33 C WBGT. Simple cognitive tasks are less sensitive to the effects of thermal stress.

Studies in this area have employed a variety of temperature scales (e.g. effective temperature, WBGT, dry bulb temperature); have required the performance of multiple tasks that may draw from different cognitive resources; were often ambiguous about task emphasis, and rarely obtained a baseline of subject performance.

While many past surveys of the literature have focused on the first two difficulties mentioned above (i.e. temperature index and task type), the issues of task emphasis and individual differences in mental capacity have not been properly dealt with in the literature.

The idea of equating baselines is not absent from the literature on attention in general. Past studies have employed this important methodology (e.g. Somberg and Salthouse 1982). A close analogy to equating the baselines of performance is found in the concept of training. If subjects are trained to a specified level of proficiency, one could argue that the task is equally difficult for all participants. This, however, is not persuasive. The method of extended practice has been criticized as an invalid method of equating. Participants will invariably differ in the time taken to achieve a threshold of performance, and this difference in the amount of practice may exert significantly on the ability to detect overall differences in performance. Any differences may, in fact, be due to differences in the level of automaticity. Further, it has been argued that heat stress limits based as they currently are, upon physiological responses to stress, ignore the true goal of productive and efficient system performance. Continuing exposure after behavioral performance and efficiency begin to fail, but before physiological limits are reached, is not appropriate for system performance and efficiency (Hancock and Vasmatzidis 1989).

7.6.2 Reducing Heat Stress

The Second Industrial Revolution is bringing about a transition in the type of work being performed. Work activities, in industrial and military settings, are becoming more cognitively demanding. Therefore, the purely physiological approach to environmental stress is no longer sufficient. A cognitive performance approach to evaluating the interface of the worker and the environment is required.

Reduction of stress due to heat in industrial and military settings can be accomplished through a variety of mechanisms. The most effective approach is to reduce heat at the source or to lower, where possible, the environmental temperature and humidity. The use of air conditioners, fans, and/or dehumidifiers can accomplish heat reduction at the source. Along the path from the source to the individual, shielding may be used. At the individual level, interventions could include: loose-fitting clothes, reduction in work rate, enforced rest breaks with provide drinking water and other fluids, and performing outdoor work at cooler times of the day.

Workers should be trained in proper work habits for work in hot environments. Use of equipment such as air-cooled, water-cooled and ice-bag vests can ameliorate the effects of heat.

In terms of heat exposure limits put forth by OSHA (Occupational Safety and Health Administration), one must look at the air velocity, the temperature in WBGT units, and the workload of the constituent tasks.

7.7 CONCLUSION

As industrial work becomes more cognitive in nature and less physical, safety issues need to evolve to account for the hazards that environmental stress may impose on the workforce. There is a need for universal standards in regard to environmental stressors for different categories of work. Heat, in particular, is a complicated issue, and when multiple stressors are imposed, things become quite complex. Data concerning the effect of environmental heat on human cognitive performance in military and industrial settings is one clear future need. A further area of fruitful investigation would be the derivation of predictive models of human behavior under stress. Such models could be quite valuable in the simulation of performance and human reliability in both extreme and more common work environments.

Once the decrement in performance due to heat is observed, one must determine the point of decrement onset by testing at a variety of temperatures. Further aspects of performance can also be investigated, such as the ability to allocate attention in varying emphasis levels. By requiring participants to switch attention between two tasks, it can be determined whether heat affects the ability to ignore irrelevant information, attend solely to one task, switch the emphasis level, and equally to share attention among two equally discriminable tasks.

By examining behaviorally based measures rather than physiologically based measures, heat standards and predictive models can be developed that focus on behavioral performance. In this way, exposure would not continue past the point of performance breakdown, which may occur before physiological system limits are reached. This area of investigation has come a long way, but future efforts will no doubt lead to more useful findings concerning heat and human performance.

REFERENCES

Kahneman, D. (1973). *Attention and effort.* Englewood Cliffs, NJ: Prentice Hall.

Hancock, P. A., & Vasmatzidis, I. (1989). Human occupational and performance limits under stress: The thermal environment as a prototypical example. *Ergonomics, 41*(8), 1169–1191.

Proctor, R. W., & Van Zandt, T. (1994). *Human factors in simple and complex systems.* Boston, MA: Allyn & Bacon.

Ramsey, J. D. (1995). Task performance in heat: A review. *Ergonomics, 38*(1), 154–165.

Somberg, B. L., & Salthouse, T. A. (1982). Divided attention abilities in young and old adults. *Journal of Experimental Psychology: Human Perception and Performance, 8*(5), 651–663.

Yerkes, R. M., & Dodson, J. D. (1908). The relation of strength of stimulus to rapidity of habit formation. *Journal of Comparative Neurology and Psychology, 18*(5), 459–482.

FURTHER READING

Brimicombe, C., Lo, C. H. B., Pappenberger, F., Di Napoli, C., Maciel, P., Quintino, T., … & Cloke, H. L. (2023). Wet bulb globe temperature: Indicating extreme heat risk on a global grid. *GeoHealth*, *7*(2), e2022GH000701.

Lu, Y. C., & Romps, D. M. (2023). Is a wet-bulb temperature of 35°C the correct threshold for human survivability?. *Environmental Research Letters*, *18*(9), 094021.

Ma, Q., Chen, Y., & Ionita, M. (2024). European summer wet-bulb temperature: spatio-temporal variations and potential drivers. *Journal of Climate*, *37*(6), 2059–2080.

Patton, E., Li, W., Ward, A., & Doyle, M. (2024). Wet bulb globe temperature from climate model outputs: a method for projecting hourly site-specific values and trends. *International Journal of Biometeorology*, *68*(12), 2663–2676.

Setyawan, A., Susilawati, S., Sutandi, T., & Najmudin, H. (2021). Performance of air conditioning unit under constant outdoor wet-bulb temperature and varied dry-bulb temperature. *International Journal of Heat & Technology*, *39*(5), pp. 1483–1490.

8 Work in Extreme Environments: Current and Future Directions

M. Kolich

8.1 INTRODUCTION

In two previous reviews, work in extreme environments was described from two perspectives. They were (a) physiological fundamentals and (b) effects on performance. With this as a base, work in extreme environmental conditions is an active and worthwhile area of future research. Examples include (a) protecting the worker from environmental extremes, particularly with respect to protective clothing, (b) air pollution and the indoor climate, (c) the development of more robust stress limits, (d) the design of appliances, (e) work in nontraditional environments (automotive, deep sea, outer space, etc.), and (f) alternative methods of evaluation (e.g. meta-analysis). This chapter emphasizes a few of these research opportunities.

DOI: 10.1201/9781003512257-8

8.2 DESIGN OF PROTECTIVE CLOTHING

8.2.1 Protective Clothing for Heat Stress

Specific to heat stress, research should be focused on protecting the worker through (1) specifying safe work–rest cycles, (2) designing "cool spots" through the use of air conditioning, fans, or dehumidifiers—a "cool spot" is typically constructed adjacent to the work area in which rest breaks are taken, and (3) providing protective clothing (liquid-cooled suits or ice-cooled vests).

In some situations, it is not possible to provide workers with liquid- or air-cooling garments because they interfere with task performance. In this context, Constable *et al.* (1994) investigated the effectiveness of using a cooling garment only during rest periods on the work capacity of subjects. They concluded that cooling during rest was effective, mostly in the first few minutes of resting, and would be even more effective if work-rest cycles were reduced to less than 30 minutes. More of this type of research is necessary, particularly as it relates to firefighting, hazardous waste disposal, and military applications.

While protective clothing is being continually improved and lightened, the requirement for adequate environmental protection is generally contradictory to the desire for adequate ventilation. For example, the requirement to meet National Fire Prevention Association (NFPA) standards for protective clothing has led to firefighting clothing ensembles that create greater heat stress for the individual than protective ensembles that were used before this standard was introduced in 1987 (Smith *et al.* 1995). These competing requirements need to be investigated.

Related to this topic, some research has been conducted concerning heat and moisture transfer through different kinds of fabrics (Wang and Yasuda 1991). In these studies, hot plates with moisture vapor production are used to mimic the behavior of heat and moisture within the clothing system of a dressed human. However, these studies have not been able to elucidate all aspects of heat and moisture transfer in the subjects under real wearing conditions. Therefore, systematic studies of the human body-clothing environment are still necessary, with subjects under various environmental and clothing conditions.

8.2.2 Protective Clothing for Cold Stress

Proper clothing is very important for reducing the adverse effects of cold on comfort, health, and performance. At present, there are very few established practices. One exception involves glove usage. If, however, finger dexterity is required, regular gloves may not be feasible. Riley and Cochran (1984) have performed some preliminary work concerning the use of partial gloves (these are gloves with the extreme and middle phalanges of the fingers exposed). It appears as though these gloves provide at least some carry-over protection to the exposed portions of the fingers under cool (if not cold) conditions. This work needs to be extended before it can be recommended for work in cold environments.

As proposed by Lockhart and Kiess (1971), one way to minimize the reduction of hand–skin temperature is to use infrared auxiliary heaters. Unfortunately, the use of such auxiliary heaters is not feasible in some situations (such as in some construction or logging operations). Other alternatives need to be explored.

There are times when exposure to cold exceeds reasonable tolerance levels. When such tolerance levels are exceeded, individuals should go to a location (such as a warm room) where rewarming of the body or hands can take place. However, there are still questions about how such rewarming should be carried out. In citing the results of various relevant investigations, for example, Enander (1984) reports that the effects of intermittent warming and cooling remain unclear. The ambiguities revolve around the question of optimum schedules for doing so. This represents another opportunity for future research.

8.3 AIR POLLUTION AND THE INDOOR CLIMATE

As a result of modern industrial activity, automobile traffic and power plants, toxic gases, such as carbon monoxide, sulfur dioxide, and hydrogen sulfide, represent an important aspect of air pollution. The presence of these gases, especially carbon monoxide, reduces the affinity of hemoglobin to bind with oxygen. Ultimately, less oxygen is delivered to the working muscles. Obviously, this can be a real handicap during work.

There is an overwhelming lack of available literature pertaining to the effects of air pollution on performance in industrial settings. The reason for this is that it is unethical to expose subjects to different levels of pollution in a work or laboratory setting. This is, however, an area that deserves more attention. With the advent of new, more sophisticated logging systems, the effects of air pollution can be more easily examined. As an example, it would be valuable to determine the effect of work rate on carbon monoxide uptake, and to what extent physical fitness may influence this relationship.

A great deal of attention has been paid recently to the air quality inside buildings. The term "sick building syndrome" has been coined to describe unduly high absenteeism among the occupants of buildings with ostensibly poor air quality. Particular attention has been paid to sealed, mechanically ventilated buildings that may be either permanently or only temporarily "sick."

Modern office buildings use sophisticated air-conditioning systems to provide workers with a comfortable indoor climate to work in. To conserve energy on heating and cooling outside air, some of the air already in the building is recirculated. Under certain circumstances, this recirculation can cause the concentrations of everyday substances to reach unacceptable levels, which may, in turn, cause health complaints.

Many factors can degrade the air quality inside buildings. Despite sometimes sensationalistic reporting, the sick building syndrome would appear to have some physical basis. Research is required to elucidate these effects. Bridger (1995) has summarized the preliminary findings, which should be used as a springboard for future research, as follows:

Symptoms are more common in air-conditioned than naturally ventilated buildings.
Clerical staff are more likely than managers to suffer.
People with the most symptoms have the least perceived control over their environment.
Symptoms are more common in the afternoon than in the morning.

8.4 DEVELOPMENT OF MORE ROBUST STRESS LIMITS

While much is known, there is still more work to do. Consider, for example, the fact that international standards for work in extreme environmental conditions, specifically ISO 7243 (Wet Bulb Globe Temperature Index) (ISO 1989b) and ISO 7933 (Sweat Required) (ISO 7933), were designed with European and North American subjects, primarily for use in those continents. These standards may (a) have limited applicability in regions that use vastly different practices (due to technological differences), (b) be culturally incompatible, or (c) ignore physiological and anatomical differences. In this context, McNeill and Parsons (1999) evaluated the usability and validity of ISO 7933 (ISO 1989a) and ISO 7243 (ISO 1989a) in terms of a simulated agricultural task in a tropical climate. They found that ISO 7243 (ISO 1989a) was valid (if slightly overprotective) and usable. ISO 7933 (ISO 1989b) underestimated sweat and evaporation rates in its prediction. The discrepancies between predicted and observed results were attributed primarily to the calculations related to clothing in the standard. Furthermore, ISO 7933 (ISO 1989b) is unusable without a computer. In regions where access to such technology may be limited, a simpler method of presentation is required.

Along the same lines, Hancock and Vasmatzidis (1998) challenge the contemporary stress limits for workers exposed to adverse thermal conditions. They argue that the basis for worker protection must shift from physiological concerns to information-processing evaluation. Thus, ergonomics, with its study of behavioral performance response, must play a much greater, if not primary, role in the development of the exposure criteria than it has done previously. Hancock and Vasmatzidis (1998) feel that exposure criteria should be developed to integrate an understanding of behavioral response to work demands with physiological adjustment to environmental conditions.

8.5 DESIGN OF APPLIANCES

Siekmann (1990) has carried out some interesting work to specify safe surface temperatures of appliances (safe in the sense that the risk of burns is reduced if the user accidentally touches the appliance). This work is interesting and useful because it leads to the drafting of very clear specifications and guidelines. However, it could be extended to look at novel ways of providing users with feedback about the state of an appliance. For example, should a domestic kettle change color as it heats up to signal that it is full of boiling water? A more general question is, "can feedback provided by appliances about their state be augmented in useful ways to enhance safety?"

8.6 WORK IN NONTRADITIONAL ENVIRONMENTS

The environments in which humans live and work will become even more extreme as technology advances. Three examples include: the deep sea, outer space, and the automotive environment. The deep sea is totally dark, extremely cold, and subject to intense pressure, while the outer space environment has no gravity or air and is subject to heavy radiation. An artificial environment will be required for humans to function under these conditions. The automotive environment (in terms of heating, ventilating, and air-conditioning systems and seats) is another burgeoning topic (Brooks and Parsons 1999).

8.7 ALTERNATIVE METHODS AND APPROACHES OF EVALUATION

The extent of human heat and cold strain depends on the interaction of variables that describe the person, the environment, the activity, and the clothing. Clearly, it is impractical to investigate all variables empirically for every human activity, so an alternative method of evaluation is essential: mathematical modeling provides this alternative. As an example, Pilcher *et al.* (2002) used a meta-analytic technique to provide a comprehensive, quantitative analysis of the effects of temperature exposure on performance. Meta-analytic reviews, because of their mathematical nature, have several statistical advantages over the more common narrative reviews. Their results indicate that hot and cold temperature exposure has a negative impact on performance and that other variables (e.g. length of exposure to the temperature or task duration) may modify this relationship. The question of ethics and morality was briefly addressed in the context of air pollution. Rodahl (1994) recounted how Norway was considering lowering the upper limit for SO_2 concentration of the working atmosphere from 2 p.p.m. to 1 or even 0.5 p.p.m. A study showed that lowering the upper limit of SO_2 concentration would mean that the plant under study would have to close. Ultimately, the upper limit of SO_2 concentration was not lowered. This may not necessarily have been the best decision for the workers involved. In this case, it seems as though money was put ahead of human well-being. As with other branches of ergonomics, it is necessary to develop a sensitivity to the cost–benefit implications and practicalities associated with these types of findings. There are, in general, very few convincing studies of the cost-effectiveness of interventions associated with the ergonomic aspects of work in extreme environments. While a participatory approach may be the most effective way to determine how the research will be perceived, this requires a radical shift from current practices—a shift from measurement to an analysis of the functioning of the entire system.

REFERENCES

Bridger, R. S. (1995). *Introduction to ergonomics (Chapter 9).* New York: McGraw-Hill.

Brooks, J. E., & Parsons, K. C. (1999). An ergonomics investigation into human thermal comfort using an automobile seat heated with encapsulated carbonized fabric (ECF). *Ergonomics*, *42*(5), 661–673.

Constable, S. H., Bishop, P. A., Nunneley, S. A., & Chen, T. (1994). Intermittent microclimate cooling during rest increases work capacity and reduces heat stress. *Ergonomics*, *37*(2), 277–285.

Enander, A. (1984). Performance and sensory aspects of work in cold conditions: A review. *Ergonomics*, *27*(4), 365–378.

Hancock, P. A., & Vasmatzidis, I. (1998). Human occupational and performance limits under stress: The thermal environment as a prototypical example. *Ergonomics*, *41*(8), 1169–1191.

International Standards Organization (ISO). (1989a). *ISO 7243: Hot environments - Estimation of heat stress on working man based on the WBGT-index (Wet Bulb Globe Temperature).* Geneva: ISO.

International Standards Organization (ISO). (1989b). *ISO 7933: Hot environments - Analytical determination and interpretation of thermal stress using calculation of required sweat rate.* Geneva: ISO.

Lockhart, J. M., & Kiess, H. O. (1971). Auxiliary heating of the hands during cold exposure and manual performance. *Human Factors*, *13*(6), 457–465.

McNeill, M. B., & Parsons, K. C. (1999). Appropriateness of international heat stress standards for use in tropical agricultural environments. *Ergonomics*, *42*(6), 779–797.

Pilcher, J. J., Nadler, E., & Busch, C. (2002). Effects of hot and cold temperature exposure on performance: A meta-analytic review. *Ergonomics*, *45*(10), 682–698.

Riley, M. W., & Cochran, D. J. (1984). Dexterity performance and reduced ambient temperature. *Human Factors*, *26*(2), 207–214.

Rodahl, K. (1994). *Stress monitoring in the workplace.* London: Lewis Publishers.

Siekmann, H. (1990). Recommended maximum temperatures for touchable surfaces. *Applied Ergonomics*, *21*(2), 69–73.

Smith, D. L., Petruzzello, S. J., Kramer, J. M., Warner, S. E., Bone, B. G., & Misner, J. E. (1995). Selected physiological and psychobiological responses to physical activity in different configurations of firefighting gear. *Ergonomics*, *38*(10), 2065–2077.

Wang, J. H., & Yasuda, H. (1991). Dynamic water vapor and heat transport through layered fabrics. Part I: Effect of surface modification. *Textile Research Journal*, *61*(1), 10–20.

FURTHER READING

Khovalyg, D., Chatterjee, A., Sellers, A. J., & van Marken Lichtenbelt, W. (2023). Behavioral adaptations to cold environments: A comparative study of active nomadic and modern sedentary lifestyles. *Building and Environment*, *243*, 110664.

International Standards Organization (ISO) (2017). *ISO 7243 Ergonomics of the thermal environment - Assessment of heat stress using the WBGT (wet bulb globe temperature) index.* Geneva: ISO.

International Standards Organization (ISO) (2023). *ISO 7933: Ergonomics of the thermal environment - Analytical determination and interpretation of heat stress using calculation of the predicted heat strain*. Geneva: ISO.

Zhang, S., Zhu, N., & Lv, S. (2021). Human response and productivity in hot environments with directed thermal radiation. *Building and Environment, 187,* 107408.

Yermakova, I. I., Potter, A. W., Raimundo, A. M., Xu, X., Hancock, J. W., & Oliveira, A. V. M. (2022). Use of thermoregulatory models to evaluate heat stress in industrial environments. *International Journal of Environmental Research and Public Health, 19*(13), 7950.

Performance Effects of High G Environments

9

W. B. Albery

9.1 INTRODUCTION

Since man first flew in a powered aircraft in 1903, he has had to perform flying tasks in a 1 G environment—macrogravity. Macrogravity poses problems to the human physiologically, ergonomically, and cognitively. Physiologically, the human loses eye-level blood pressure at 22 mmHg/G as G increases from 1.0 to 9.0, for example, in the F-16 aircraft. To maintain consciousness and perform in the high G environment, humans must be protected against the effects of high G. Ergonomically, human gets heavier as G increases. A 200 lb man weighs 1800 lb at 9 G! Arms, legs, and head are heavier, and tasks such as flipping switches on the instrument panel require trajectory strategy and practice with a 10 lb arm that now weighs 90 lb. Cognitively, the loss of eye-level blood pressure and reduced eye-brain-level arterial oxygen saturation (80%) relate to flying performance decrement. What can be done to counter these effects of high G? Modern air forces have developed countermeasures to these performance effects of high G environments.

DOI: 10.1201/9781003512257-9

9.2 FUNDAMENTALS OF THE HIGH G ENVIRONMENT

The body axes for sustained acceleration are shown in Figure 9.1. The inertial resultants of body acceleration are listed in Table 9.1. Centrifugal acceleration is expressed in terms of angular velocity and the radius of turn:

$$G\left(w^2 r\right)/g,$$

where w = angular velocity in radians/s and G = unitless measure of acceleration, r = radius of turn (m) and g = gravitational constant (9.8 ms^{-2}). A typical human centrifuge with a 20-ft radius (6.1 m) can generate 9 G by turning at 38 revolutions/min, which is ~4 radians s^{-1}.

The easiest axis for the human to tolerate sustained acceleration is in the chest-to-back (+G*x*) direction. This is why astronauts are launched on their backs. Volunteers have been exposed up to +16.5 G*x* unprotected for brief periods. Subjects completely immersed in water have endured >30 G (von Gierke *et al.* 1991). The least tolerable direction is the -G*z* or eyeballs-down direction. Some pilots have been exposed to brief periods of -G*z*, but the increased blood

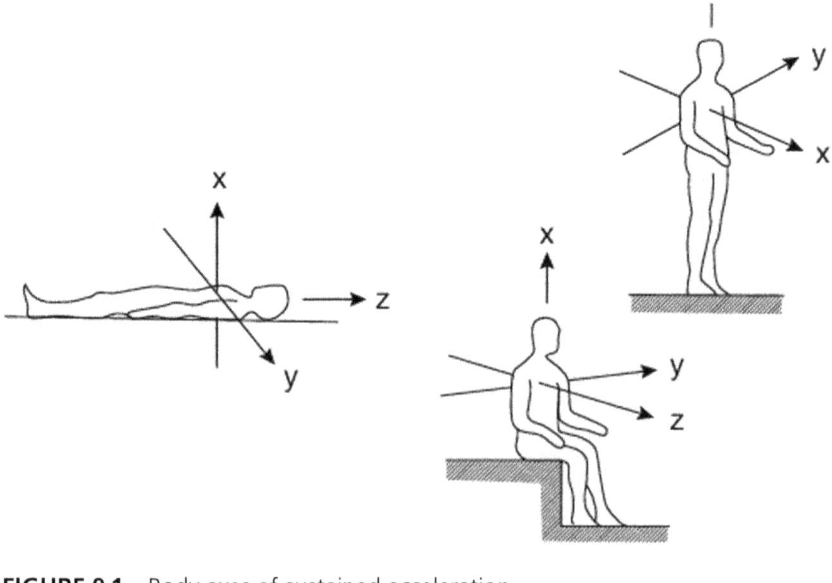

FIGURE 9.1 Body axes of sustained acceleration.

TABLE 9.1 Nomenclature for Inertial Resultant of Body Acceleration

LINEAR MOTION	*PHYSIOLOGIC DESCRIPTIVE*	*PHYSIOLOGIC STANDARD*	*VERNACULAR DESCRIPTIVE*
Forward	Chest-to-back G	+G*x*	Eyeballs-in
Backward	Back-to-chest G	-G*x*	Eyeballs-out
Upward	Positive G	+G*z*	Eyeballs-down
Downward	Negative G	-G*z*	Eyeballs-up
To right	Left lateral G	-G*y*	Eyeballs-left
To left	Right lateral G	+G*y*	Eyeballs-right

pressure at the eye-brain level is very uncomfortable. The lateral direction (±G*y*) is less tolerable than ±G*x* exposures. Most high G aircraft routinely expose the pilot(s) to +G*z*. Man, protected and straining, can achieve +9 G*z* for short periods and perform complex piloting tasks.

9.3 PHYSIOLOGICAL EFFECTS OF HIGH G

The high G environment is also characterized by its effect on human physiology. The accepted model of human G tolerance is the hydrostatic column model of man (Burton and Whinnery 1996). In this model, the eye-heart distance results in a change in blood pressure measured at heart level and at eye level (Figure 9.2.).

It is shown that, as one sits in the normal 1 G environment, there is a differential in the eye-to-heart blood pressure. This is illustrated in the seated pilot at the left. The effect of additional +G*z* is illustrated, where the pilot's neck is elongated. Since each G costs 22 mmHg blood pressure, at 6 G the pilot has −12 mmHg blood pressure at eye level. This −12 mmHg blood pressure must be raised >0 to keep the eyes and brain perfused. The loss of eye-level blood pressure is typically countered by three protective techniques:

- Cardiac compensation: the baroceptors in the circulatory system sense a reduced blood pressure and signal an increase in heart rate. The increased heart rate increases blood pressure.
- Straining maneuver: a self-initiated strain against a closed glottis that raises intrathoracic pressure, which in turn increases eye-level blood pressure.
- Anti-G suit/positive pressure breathing: artificial means of raising one's blood pressure; the anti-G suit inflates and increases pressure

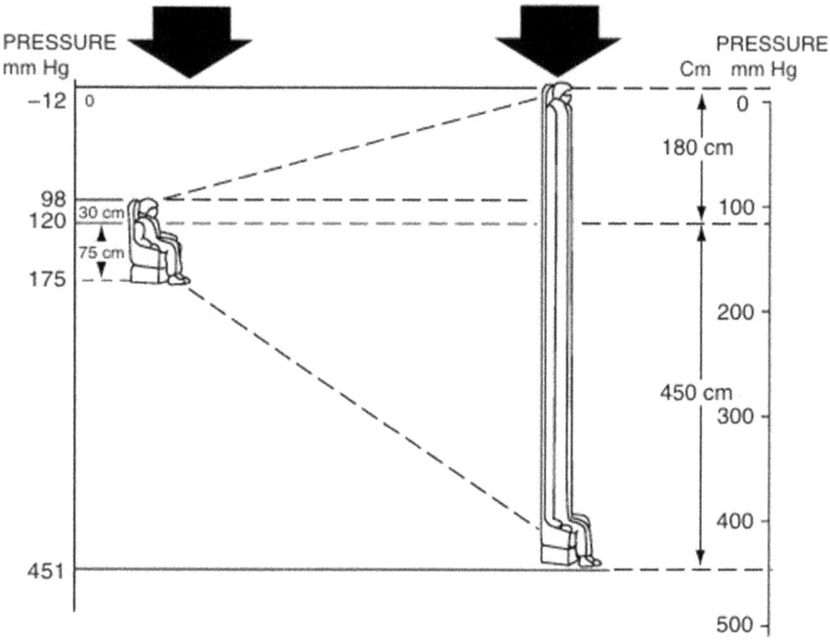

FIGURE 9.2 Hydrostatic column equivalents (arterial side).

to the legs and abdomen. This increased pressure aids in venous return to the heart, which improves G tolerance. The US Air Force has developed a positive pressure breathing system that inflates the pilot's lungs as a function of increasing G (12 mmHg/G to a maximum of 60 mmHg). This system, called COMBAT EDGE, has helped protect humans to +12 G*z* for brief periods.

9.4 ERGONOMIC EFFECTS OF THE HIGH G ENVIRONMENT

The pilot becomes heavier in the seat as G increases. Arms, legs, torso, and head all become heavier, and with the same motor control and musculoskeletal systems used at normal 1 G, man initially has problems controlling movements of the head, arms, and legs at high G.

9.4.1 Ability to Move under Increased G

Increased gravitational fields increase the weight of body parts. Body parts can become elongated or compressed under the G vector; this can affect the shape and function of the soft internal organs, including the heart, lungs, kidneys, liver, etc. Higher muscle forces are required to keep the head, torso, and limbs in desired positions. At G forces of ~ +2 *Gz* there is increased pressure on the buttocks, drooping of the face, and noticeably increased weight of all body parts; at this level of G force it is difficult to raise oneself, and at +3-4 *Gz* it is nearly impossible. Experiments were conducted in the Wright-Patterson AFB (WPAFB) Dynamic Environment Simulator (DES) centrifuge, where subjects arose from a seated position and performed a whole-body jump at *Gz* levels up to 1.8 *Gz* as the centrifuge rotated. Although most subjects had no problem standing up at 1.8 *Gz*, jumping and leaving one's feet was very difficult (Albery and Woolford 1997). Above +3-4 *Gz*, controlled motions require greater effort, adaptation, and learning to offset loss of fine motor control. While seated in a high-performance aircraft, one typically cannot raise the arm at forces >8 *Gz* or legs at >3 *Gz*. Head pitching is difficult at >4 *Gz*, and some individuals who get their heads pitched forward at high *Gz* (>6 G) are unable to right themselves in the seat until the acceleration is unloaded. The hand can be raised slightly at 25 *Gz*. One can barely slide the feet on a floor >5 *Gz*. Speech is severely affected, yet possible, up to +9 *Gz* if the operator is utilizing protective techniques properly (see above).

9.4.2 Arm Trajectory at High G

When reaching for a switch on the instrument panel at high G, the tendency of pilot is to complete the movement below the switch because the arm is now heavier due to increased +*Gz*. The pilot learns after several attempts to initiate the trajectory of the arm-hand at a location above the switch. A re-calibration of gross motor movement occurs during long exposure to even low-level acceleration (Albery and Woolford 1997).

9.5 COGNITIVE EFFECTS OF HIGH G

Very rarely does high-sustained acceleration affect only one factor, such as speech, vision, or manual dexterity. Task performance is manually measured through subjective reports, physiological, and behavioral techniques.

Cognitive performance is significantly affected when eye-level arterial oxygen saturation is <80%. Percent arterial oxygen saturation (%S_a O_2), as measured with a pulse oximeter, can drop to <80% when the ventilation/perfusion ratio is not evenly matched. This can occur when subjects are exposed to high G. Complex task performance has been shown to deteriorate when %S_aO_2 (Albery and Chelette 1998). Percent cerebral oxygen saturation (%rO_2) is also used as a physiological measure in centrifuge research (Chelette *et al.* 1998).

9.5.1 Characterizing Human Performance at High G: Exposures Up to 3 Minutes

In studies at the Air Force Research Laboratory, WPAFB, human performance during simulated aerial combat sorties has been characterized via "performance surfaces" (Figure 9.3). These 3-D projections allow the observer to look at the cognitive and physiological performance of the subject as a function of brief exposures to sustained acceleration (+G*z*).

In Figure 9.3, the performance task score is shown on the vertical axis, time spent at G*z*>8 is shown on the horizontal axis, and the physiological variable, minimum cerebral oxygen saturation (min rO_2), is shown on the other horizontal axis. One can see from this performance surface that the best performance in terms of score was when the subjects spent only 10 s at 8 G and when rO_2 was at its highest, 65%. Chelette *et al.* (1998) compared male and female performance during a 3-min simulated air-to-air combat situation that exposed subjects to 9 G*z*. They found no significant differences between male and female flying performance.

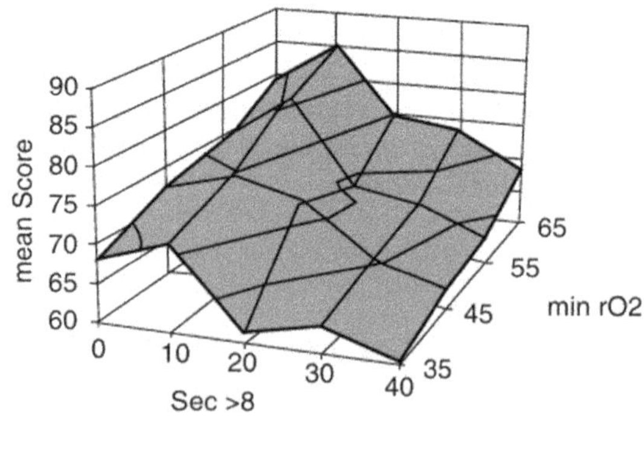

FIGURE 9.3 Performance surface.

9.5.2 Characterizing Human Performance at High G: Exposures from 3 to 11 Minutes

Some pilots and many centrifuge subjects have endured G exposures that included peaks to 9 G for several minutes. While evaluating new protective garments, humans have endured 10 *Gz* continuously for 3 min. Subjects performing a dual-task simulated aerial combat maneuver in the DES centrifuge endured up to 27 peak exposures to 9 G over 11 min and still maintain excellent tracking and secondary task performance. This exposure is much longer than what a typical dogfight would probably last, but it illustrates the enhanced protection of humans wearing new COMBAT EDGE-type G protection garments.

9.5.3 Effect of High G on Orientation Perception

Sustained acceleration (G>1) can also have a profound effect on human perception. Some of these effects are discussed in the previous chapter. It is believed that the otolith organs of the vestibular system are gravity transducers (graviceptors) and, as such, play an important role in human perception of motion, especially in the absence of visual cues.

9.5.3.1 Somatogravic Illusion: High G Pitch-Up Sensation

Sustained acceleration experienced by pilots during catapult launching of Navy aircraft off the decks of aircraft carriers has been implicated as a contributing factor to aircraft mishaps. The catapult launch generates a + *Gx* acceleration on the pilot, which can elicit a somatogravic illusion (a false sensation of body tilt that results from perceiving as vertical the direction of a non-vertical gravito-inertial force). Pilots of carrier-launched aircraft need to be especially wary of the somatogravic illusion. These pilots experience pulse accelerations lasting 2–4 s and generating peak inertial forces of +3-5 *Gz*. Although the major acceleration is over quickly, the resulting illusion of nose-high pitch can persist for >30 s afterward, resulting in a particularly hazardous situation for the pilot unaware of the phenomenon. Navy scientists reproduced on the centrifuge the pitch-up sensation experienced by pilots during catapult launches, and warnings were issued to all pilots taking off from aircraft carriers.

9.5.3.2 G-excess Illusion

The G-excess illusion is a false or exaggerated sensation of body tilt that can occur when the G environment is sustained at >1 G. This illusion has been implicated in a number of low-level turning and looking mishaps in the USAF. It is believed that the otoliths "over report" a roll or pitch when the pilot and the aircraft are in a >1 G maneuver. This over reporting occurs because the otoliths are "tugged" more than usual when G > 1, and this is perceived by the pilot as an increased roll or pitch attitude. If a pilot experiences the G-excess illusion, the tendency is to compensate for the illusion, which can cause overbanking, loss of altitude, and even controlled flight into the terrain. Chelette *et al.* (1998) duplicated the G-excess illusion on the WPAFB DES centrifuge (Albery 1997). Rolling and/or pitching the head while in an excess G environment (1) can cause an illusory sensation of vehicle tilt. This illusion occurs in the pitch axis if the head is forward, but translates to the roll axis as the head is turned toward one shoulder.

9.5.4 Time and Mass Estimation in Macrogravity

Time, mass, and reaction time perception experiments have been conducted under high levels of G stress (Albery *et al.* 1997). G levels up to and including +8 G*z* affect time perception. The longer the time estimation, such as >15 s, the greater the error. Subjects tended to underestimate the length of longer tasks, as if they wanted to finish the task early. Likewise, for mass discrimination, it was found that performance is impaired during increased +G*z*. The ability to discriminate between the heaviness of several objects became more difficult under G stress. In several studies involving choice reaction time (more than one possible selection) in acceleration stress, it was found that reaction times actually decrease under G stress. This effect of improved reaction time performance under acceleration can be possibly attributed to the alerting cue phenomenon (subjects are more "on their toes" when exposed to short periods of sustained acceleration).

9.6 CONCLUSIONS

Most humans can tolerate and perform in G levels up to 4 G for tens of seconds without straining and unprotected. Above 4–6 G*z*, the loss of eye-level blood pressure must be countered for the pilot to remain conscious. Performance is degraded. By straining and raising one's eye-level blood pressure, unprotected pilots can be trained to endure 9 Gz and to perform complex tasks for brief periods of exposure (10–15 s).

Using anti-G protective pants, vest, and positive pressure breathing, pilots can be trained to endure 19 Gz and perform complex flying tasks for minutes at a time. Human task performance tends to decrease under high G as a function of decreasing eye-level blood pressure and reduced arterial oxygen saturation, which directly affect cognition.

No gender differences have been found in male-female flying performance in the high G environment, when the subject/pilot is well protected.

High G can have profound effects on vision, orientation perception, and time and mass perception.

REFERENCES

Albery, W. B. (1997). Evaluation of six G protection ensembles during 5 to 9 G simulated aerial combat maneuvers. *SAFE Journal*, *27*(2), 92–99.

Albery, W., & Chelette, T. (1998). Effect of G-suit type on cognitive function. *Aviation Space and Environmental Medicine*, *69*, 474–479.

Albery, W., & Woolford, B. (1997). Design for macrogravity and microgravity experiments. In G. Salvendy (Ed.), *Handbook of human factors and ergonomics* (2nd ed., pp. 935–963). Wiley.

Burton, R., & Whinnery, J. (1996). Biodynamics: Sustained acceleration. In R. DeHart (Ed.), *Fundamentals of aerospace medicine* (2nd ed., pp. 201–260). Williams & Wilkins.

Chelette, T., Albery, W., Esken, R., & Tripp, L. (1998). Female exposure to high G: Performance of simulated flight after 24 hours of sleep deprivation. *Aviation Space and Environmental Medicine*, *69*, 862–868.

Von Gierke, H., McCloskey, K., & Albery, W. (1991). Military performance in sustained acceleration and vibration environments. In R. Gal & D. Mangelsdorff (Eds.), *Handbook of military psychology* (pp. 335–364). Wiley.

FURTHER READING

Lehman-Wilzig, S. N., & Lehman-Wilzig, S. N. (2021). Mathematics, Philosophy, Physics and Cosmology. *Virtuality and Humanity: Virtual Practice and Its Evolution from Pre-History to the 21st Century*, 73–101. Springer.

Mouloua, M., French, J., Smither, J. A., & Kennedy, R. S. (2023). Mitigation of Motion Sickness Symptoms by Adaptive Perceptual Learning: Implications for

Space and Cyber Environments. In *Human factors in simulation and training* (pp. 179–208). CRC Press.

Wenzel, E. M., & Godfroy-Cooper, M. (2021). Advanced multimodal solutions for information presentation (No. NASA/TM-20210017507).

Xing, C., Gao, Y., Wang, X., Xing, W., Liu, Y., Lei, Y., … & Gao, F. (2021). Cuff-method thigh arterial occlusion counteracts cerebral hypoperfusion against the push: Pull effect in humans. *Frontiers in Physiology*, *12*, 672351.

Zhao, S., Ke, P., & Duan, C. (2024, August). Prediction and improvement of human heat stress based on numerical model. In *CSAA/IET International Conference on Aircraft Utility Systems (AUS 2024)* (Vol. *2024*, pp. 2221–2226). IET.

Tolerance to Sustained $+G_z$ Acceleration

10

R. R. Burton

10.1 INTRODUCTION

The environments produced by accelerating a human body are grouped into three main categories: (1) transient, (2) sustained and (3) chronic. Acceleration takes place when the velocity of a body changes either in magnitude or direction. Changes in acceleration that last 1 or 2 s and do not achieve steady-state are termed transient, abrupt or impact. Sustained acceleration is usually a steady-state environment with short periods of changing accelerations at the beginning (onset) and ending (offset) of the acceleration environment. The duration of this exposure is usually 1 h. Chronic acceleration is a steady-state exposure that lasts for several hours, days or even years. Not surprisingly with these types of definitions that rely on the duration of exposure, there are no clear distinctions where transient acceleration ends and sustained acceleration begins. The same can be said about sustained and chronic accelerations. Because of the very short duration of transient acceleration, the environment is developed using sleds or drop towers. This chapter deals with sustained acceleration that uses the centrifuge to provide the acceleration, as does chronic acceleration studies.

DOI: 10.1201/9781003512257-10

Reasons for studying these types of accelerations are to develop information on human effects as they primarily relate to the following activities: (1) transient for impact injuries, (2) sustained for pilot tolerance during fighter aircraft maneuvers and (3) chronic for long-term space flight.

For more information on transient acceleration, see Brinkley and Raddin (1996) and for sustained and chronic acceleration, see Burton and Whinnery (1996) and Burton and Smith (1996).

10.2 PHYSICAL PRINCIPLES

The human, in a sustained acceleration environment, physiologically responds to the inertial force generated by the acceleration. The forces of acceleration (a) are described by Isaac Newton's three laws of motion. The first law states that a body remains at rest or in motion in a straight line until acted upon by a force. The second law of motion defines the force (F) required to overcome the first law, i.e. $F = \text{mass}\,(m) \times \text{acceleration}\,(a)$. The third law states that for every action there is an equal reaction, thus establishing the physical principles of acceleration in circular motion as generated by a centrifuge, i.e. centripetal (acceleration) force (a) establishes an equal centrifugal (inertial) force (G). Thus, a centrifuge or an inside loop of an aircraft produces the same acceleration with an equal inertial force acting in the opposite direction.

Equation 10.1 predicts the centripetal acceleration resulting from the motion involved:

$$a = rv^2 \tag{10.1}$$

where a is centripetal acceleration $\left(\text{ms}^{-2}\right)$, v^2 is radians s^{-1} and r is the radius of the circle (m).

Radian degree tables are available but for simple calculations, $360^\circ = 6.28$ radians. The product of equation 10.1 is "a" $\left(\text{ms}^{-2}\right)$ that, for convenience, is divided by Earth's gravitational constant $\left(9.81\ \text{ms}^{-2}\right)$ providing a value that is in multiples of Earth's gravity $(1g)$. Thus, "a" is frequently used in place of "a." Hence, $g = G$, the inertial force (G) resulting from the acceleration (g) defined in equation (10.1) as "a" is also measured and is equal.

"G" resulting from "g" is the parameter of importance in defining human tolerance to acceleration levels (intensities). The physical and physiological effects on the body come from G. Of particular importance for the human tolerating increasing levels of sustained G, is the distribution of blood within the body.

Fluid pressures within an acceleration field obey the hydrostatic pressure law (equation 10.2):

$$P_H = hdg \tag{10.2}$$

where P_H is hydrostatic pressure, h is column height, d is the specific density of the liquid and g is the level of acceleration.

Since $g = G$, equation 10.1 predicts the effect of G on arterial pressure (P_a) and blood flow within the body. Hence, it has been adapted for use as a human G-level tolerance prediction model (equation 10.3; Burton and Whinnery 1996):

$$G_{LT} = \frac{(P_a \Sigma d)}{h} \tag{10.3}$$

where G_{LT} is G-level tolerance of a person seated upright, h is eye-heart vertical distance (350 mm), P_a is mean arterial blood pressure at eye level is 100 mmHg and d is specific density of mercury (13.6) used in blood pressure measurements.

10.3 AXIAL SYSTEM NOMENCLATURE

Since equation 10.3 predicts the G level that a human can tolerate, it becomes clear that "h" (eye-heart vertical distance) plays a major role in G-level tolerance determinations. Thus, the orientation of the body in the acceleration environment must be identified when defining G-level tolerances. Thus, a standard axial system of nomenclature has been adopted to identify this relationship for all acceleration-related studies.

10.3.1 Relaxed G Tolerance

There are two definitions of human tolerance to sustained acceleration. First there is the level of acceleration that can be tolerated with a cardiovascular basis, using visual criteria, known as G-level tolerance. The other tolerance is the duration of acceleration that can be tolerated with a metabolic basis, using symptoms of fatigue G-duration tolerance. Obviously both of these tolerances have common dimensions of G level and duration, i.e. the difference is in the dimension criteria that determine the tolerance.

10.3.2 G-Level Tolerance

Natural, control or basic human *G*-level tolerance $(+G_z)$ is measured on relaxed subjects on a centrifuge with *G* onset rates of $\geq 1\text{Gs}^{-1}$ (Burton et al. 1974). This tolerance measurement is called "rapid onset relaxed" (ROR). The subject is seated upright with "*g*" directed toward the head and, of course, corresponding *G* directed toward the feet. The *G* level is increased until a predetermined tolerance criterion is reached. Several criteria exist: (1) grayout, (2) blackout or (3) loss of consciousness. The first two criteria are visual, recognized by the subject who terminates the centrifuge exposure by activating a hand brake. Grayout (also known as peripheral light loss or tunnel vision) is a dimming of the vision while blackout (central light loss) is a total loss of vision. Grayout visual symptoms occur because blood flow in peripheral regions in the retina of the eye ceases as systolic P_a at eye-level falls from a normal level of 125 mmHg to ~ 40 mmHg. Blackout occurs with eye-level $P_a = 20$ mmHg. With blackout, subjects are still conscious, thus they can hear and are aware. This visual symptom is possible because intra-ocular pressure is 20 mmHg which causes the retinal blood vessels to collapse with ~ 20 mmHgP_a in the head. Recovery from these visual symptoms is immediate upon reestablishing blood flow to the eye.

A subject loses consciousness, known as *G*-induced loss of consciousness $(G-\text{LOC})$, when blood flow to the brain ceases. Recovery from this condition is not immediate as with visual symptoms, but requires 15 – 30 s. Thus, G-LOC is an extremely dangerous condition as the aircraft remains totally without control while the pilot is recovering consciousness. *G*-level tolerance can be predicted by solving equation 10.3 and is ~ 4*G*, i.e. each *G* requires ~ 25 mmHgP_a at heart level (Burton and Whinnery 1996).

If the subject is not physically relaxed, *G* tolerance will increase, e.g. simply gripping one hand will increase it by 0.6 G. Contraction of arm, leg and abdominal muscles can increase *G* tolerance by several *G* thus forming the basis for a maneuver that is used by pilots to increase their *G* tolerance during high *G* exposures (see below).

The use of slow-onset of *G* rates (i.e. $0.1G\ \text{s}^{-1}$), known as gradual onset relaxed (GOR) exposures, defines a *G* tolerance that is ~ 1*G* higher than ROR tolerances. GOR tolerance is increased above ROR because of the activation of a physiologic reflex (baroreceptor initiated) that increases heart rate, causes arterial vessel constriction thus raising P_a. This reflex is more pronounced in fighter pilots than subjects not regularly exposed to *G*, suggesting that a type of environmental adaptation has occurred.

10.3.3 G-Duration Tolerance

G-duration tolerances for relaxed subjects are measured using a constant *G* level that continues until the subject becomes fatigued and ends the exposure by applying a hand brake that stops the centrifuge. *G*-duration tolerances for *G* levels <≠ 5*G*) are predicted using:

$$G_D = 2666e^{-1.265G} \tag{10.4}$$

where G_D is *G*-duration tolerance (min) and *G* is *G*-level exposures ≤ 5*G*.

10.4 METHODS TO INCREASE G TOLERANCE

Sustained *G* levels above those naturally tolerated by a human is an environment that is studied because it is generated by fighter planes during maneuvers that are used in air combat, i.e. aerial combat maneuvers (ACM). The ACM is a very dynamic activity as pilots rapidly change their aircraft positions using inside loops as they attack the enemy or for defensive purposes. Therefore, fighter aircraft are designed to develop very high *G* levels (i.e. 9*G*) because highly maneuverable aircraft, that can make tighter turns or make them more rapidly, have a great advantage over their adversary during ACM.

Thus, pilots are exposed to *G* levels far in excess of their natural *G*-level tolerances of ~ 4*G*. To allow pilots to tolerate these high *G* levels for considerable periods without having *G*-LOC, methods have been developed that provide protection against *G*. These methods are called "anti-*G*" or "*G* protection." However, it is important to understand that they do not truly "protect" a pilot against the effects of *G*. Anti-*G* methods do not block the effects of *G*, but only increase pilots' *G* tolerances thus exposing them to higher *G* levels and increasing the potential hazards of *G* exposures.

10.4.1 Anti-G Suit

Various types of anti-*G* suits were developed in World War II by the USA, Canada and Australia. The US-design anti-*G* suit with a few modifications is used today by fighter pilots from all nations that have high-performance aircraft. This garment is made of fire-retardant material that covers pneumatic bladders that apply pressure to the legs and supports the abdomen (Burton and Whinnery 1996). This suit is inflated with pressurized air, controlled by an anti-*G* valve at $1.5\,\text{psi}/G$ to a maximum of 10 psi. This suit increases *G* tolerance by $1.0-1.5G$ by increasing P_a and elevating the position of the heart during *G* – the latter reducing "*h*" of equation 10.3.

10.4.2 Anti-G Straining Maneuver

Also during World War II, a physical maneuver, called the M-1, was perfected in conjunction with the anti-*G* suit. M-1 uses the combination of muscle contractions and a forced exhalation against a closed or partially closed glottis (Wood et al. 1946). This maneuver can increase pressure within the lungs by as much as 100 mmHg, resulting in a similar increase in P_a. This method is now called an Anti-*G* Straining Maneuver (AGSM) that, with the increase of P_a at eyelevel of 100 mmHg, increases *G*-level tolerance by 4*G* (equation 10.3). By combining the AGSM with the anti-*G* suit, pilots can tolerate 9*G*, to match the acceleration-level capability of their aircraft (Burton and Whinnery 1996). On the other hand, performing the AGSM is extremely fatiguing to the pilots thus limiting their *G*-duration tolerance capability.

Obviously this maneuver is very important to pilots as it forms the basis for 50% of their *G* tolerance. Since it must be learned, training programs have been developed that teach pilots, on a centrifuge, how to perform an effective AGSM.

Since muscle contractions form the basis for the AGSM, physical conditioning programs have been developed for pilots to increase their muscle strength, thus performing this maneuver with less effort and slowing the onset of fatigue.

10.4.3 Positive Pressure Breathing During G (PBG)

Positive pressure breathing (PPB) systems were first developed to protect pilots flying in low-pressure environments from hypoxia. This system forces air into the lungs above the ambient environmental pressure thus increasing oxygen levels in the arterial blood. To assist in breathing against this increased pressure in the lungs, a counter-pressure is applied outside the chest at the same pressure level, using a type of inflatable vest. This PPB system has been developed recently as a G-protection method known as positive pressure breathing during G(PBG). Pressures as high as 60 mmHg are used, which require little physical effort by the pilot. Since PBG increases pressure in the lung like the AGSM and with much less effort, pilots fatigue less rapidly thus increasing their G-duration tolerance. An operational version of PBG is now commonly used internationally by pilots flying high-performance aircraft (Burton and Whinnery 1996).

10.4.4 Reclining or Pronation of Pilots

Subjects using seats that have reclined backs are placed in a supine position and exposed to $+G_x$. Subjects can also be pronated resulting in $-G_x$ exposures. These positions reduce "h" of equation 10.3 thus significantly increasing G tolerances. A G_x level tolerance as high as 15G has been realized. Although this method for increasing G-level tolerance is very effective, cockpit design complexities have prevented its use in fighter aircraft.

10.5 HIGH SUSTAINED G (HSG) TOLERANCE

Anti-G methods described above have increased G level and duration tolerances far above normal relaxed levels. This environment is known as high-sustained G (HSG; Burton et al. 1974). Advanced anti-G methods allow > 9G to be tolerated by pilots thus providing them more than adequate protection for flying 9G fighter aircraft. These high-G-level tolerances are measured using the same visual and G-LOC criteria that determine relaxed G-level tolerances (see above).

G-duration HSG tolerances are limited because of pilot fatigue that results from performing the AGSM. These tolerances are measured using continuous G exposures that can be at a constant G level or can be variable. Variable G profiles are designed to simulate ACM that are used by pilots in combat and training thus the name Simulated ACM (SACM; Burton and Shaffstall 1980). The SACM is most commonly used to measure HSG duration tolerances. G duration tolerances for a continuous and constant G profile are predicted using:

$$G_D = 14.08e^{-0.328G} \tag{10.5}$$

where G_D is as in equation 10.4 and G is G-level exposures > 5 – 9G.

This equation can also be used to approximate SACM tolerances by increasing the predicted tolerance by 20% – 30% (Burton 1998).

10.6 PERFORMANCE WITH G EXPOSURE

Pilots must have excellent awareness during ACM, known as Situation Awareness (SA). Good SA is a very high-priority item for fighter pilots. Without a doubt their most critical SA is during ACM. To preserve their SA, pilots must maintain good vision during *G* exposures. The occurrence of tunnel vision (grayout or blackout with *G* exposures will greatly reduce their performance thus anti-*G* methods are very important for pilots' SA). Of course, *G*-LOC with loss of awareness eliminates SA and all performance capability for at least 30 s.

Subject performance during exposures to increased *G* has been measured using subjects on the centrifuge. Unfortunately values obtained from these studies are difficult to evaluate relative to pilot performance during ACM (Burton et al. 1974). These studies generally show an exponential decay in motor function (i.e. target tracking) beginning at 3*G* with increasing reductions in performance to 60% – 70% at 7 G. These levels of performance reduction during HSG appear to be unrealistically high for pilots that train to achieve near perfection in their performance during ACM.

The major problem in measuring performance relevant to the fighter pilot is the complexity of the visual and cognitive requirements of the ACM while interacting with *G*. An operational ACM cannot be accurately duplicated on the centrifuge. Perhaps in the future improved visual displays and the use of Virtual Reality may provide the means to measure pilot performance more accurately in the laboratory.

REFERENCES

Brinkley, J. W., & Raddin, J. H., Jr. (1996). Biodynamics: Transient acceleration. In R. L. DeHart (Ed.), *Fundamentals in aerospace medicine* (2nd ed., pp. 163–201). Williams & Wilkins.

Burton, R. R. (1998). Mathematical models for predicting human tolerance to sustained acceleration. Presented at the *Research and Technology Agency Conference on "Computer Models for Aircrew Safety Assessment: Uses, Limitations and Requirements"*, 26–28 October, Ohio: Wright-Patterson Air Force Base.

Burton, R. R., Leverett, S. D., Jr., & Michaelson, E. D. (1974). Man at high sustained +Gz acceleration: A review. *Aerospace Medicine*, *45*, 1115–1136.

Burton, R. R., & Shaffstall, R. M. (1980). Human tolerance to aerial combat maneuvers. *Aviation Space and Environmental Medicine*, *51*, 641–648.

Burton, R. R., & Smith, A. H. (1996). Adaptation to acceleration environments. In M. J. Fregly & C. M. Blatteis (Eds.), *Handbook of physiology, section 4: Environmental physiology* (Vol. *II*, pp. 943–970). Oxford University Press.

Burton, R. R., & Whinnery, J. E. (1996). Biodynamics: Sustained acceleration. In R. L. DeHart (Ed.), *Fundamentals in aerospace medicine* (2nd ed., pp. 201–260). Williams & Wilkins.

Wood, E. H., Lambert, E. H., Baldes, E. J., & Code, C. F. (1946). Effects of acceleration in relation to aviation. *Federation Proceedings*, *5*, 327–344.

FURTHER READING

Bacevic, N., Ninkovic, M., Drvendzija, M., Vidakovic, J., Bacevic, M., & Stepanic, P. (2024b). Heart rate variability as a predictor of+ Gz tolerance during the high-G selective test. *Aerospace Medicine and Human Performance*, *95*(2), 93–100.

Copeland, K., & Whinnery, J. E. (2023b). Cerebral blood flow based computer modeling of Gz-induced effects. *Aerospace Medicine and Human Performance*, *94*(5), 409–414.

Pollock, R. D., Hodkinson, P. D., & Smith, T. G. (2021b). Oh G: The x, y and z of human physiological responses to acceleration. *Experimental Physiology*, *106*(12), 2367–2384.

Pooja, K., Kumar, A., & Dinakar, S. (2022b). A case of low G-tolerance: An approach to aeromedical evaluation and disposal. *Indian Journal of Aerospace Medicine*, *66*(1), 39–42.

Wang, J., Cheng, P., Ji, M., Xia, X., Hu, W., Wang, P., & Wang, Z. (2024b). Mechanisms and prevention of acceleration exposure-induced cardiovascular injuries. *Occupational Diseases and Environmental Medicine*, *12*(4), 340–348.

Index

Pages in *italics* refer to figures and pages in **bold** refer to tables.

www.ingramcontent.com/pod-product-compliance
Ingram Content Group UK Ltd.
Pitfield, Milton Keynes, MK11 3LW, UK
UKHW022016111225
465993UK00008B/674
* 9 7 8 1 0 3 2 8 4 1 3 1 1 *